Human Body

Written by
Pamela Jennett

Editor: Collene Dobelmann
Illustrator: Darcy Tom
Designer/Production: Moonhee Pak/Cari Helstrom
Cover Designer: Barbara Peterson
Art Director: Tom Cochrane
Project Director: Carolea Williams

Table of Contents

Introduction

Each book in the *Power Practice*™ series contains over 100 ready-to-use activity pages to provide students with skill practice. The fun activities can be used to supplement and enhance what you are teaching in your classroom. Give an activity page to students as independent class work, or send the pages home as homework to reinforce skills taught in class. An answer key is provided for quick reference.

The practical activities, charts, diagrams, and definition pages in *Human Body* supplement and enrich classroom teaching to enhance students' understanding of vocabulary, functions, and processes fundamental to the human organism. The following systems are covered in the activities:

- skeletal
- muscular
- circulatory
- respiratory
- digestive
- urinary
- lymphatic
- endocrine
- central nervous
- sensory
- reproductive

Use these ready-to-go activities to "recharge" skill review and give students the power to succeed!

Name ______________________________ Date ______________

The Human Cell

The cell is the smallest entity that still retains the properties of life. A cell can survive on its own or has the potential to do so. This diagram shows an example of an animal cell. While each type of cell differs somewhat, they all share similar structures and functions. Use the terms in the word box to label the diagram.

vacuole	Golgi body	cell membrane	cytoplasm
mitochondrion	endoplasmic reticulum	nucleus	

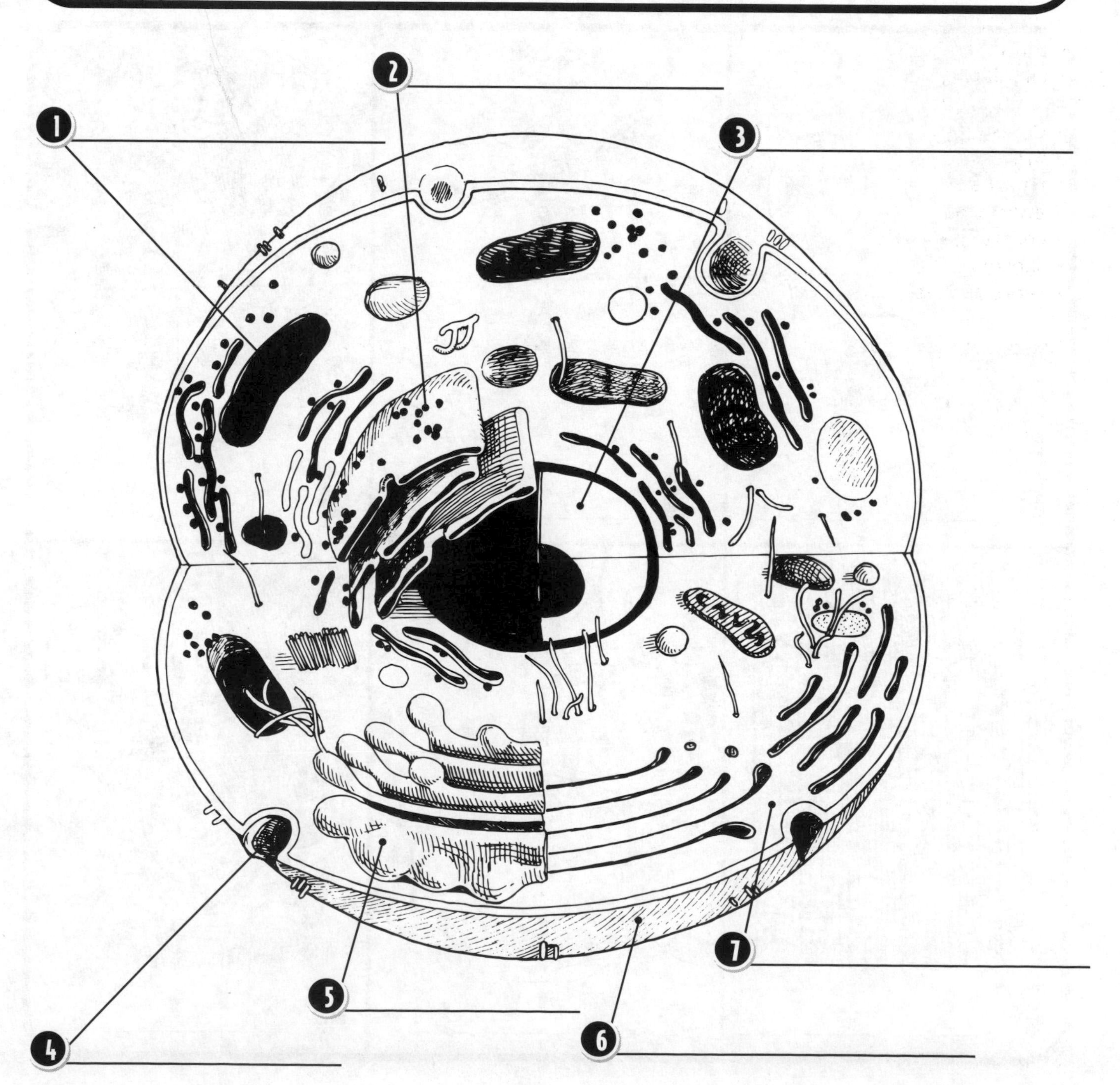

Name ______________________ Date ____________

Types of Cells

Cells differ enormously in shape, size, and activities. Yet all cells are alike in three aspects. All start out with a plasma membrane, a region of DNA, and a region of cytoplasm. Many cells have a nucleus; some do not. Use the terms in the word box to label the types of cells.

red blood cells	bone cells	cartilage cells
nerve cells	skeletal muscle cells	adipose cells

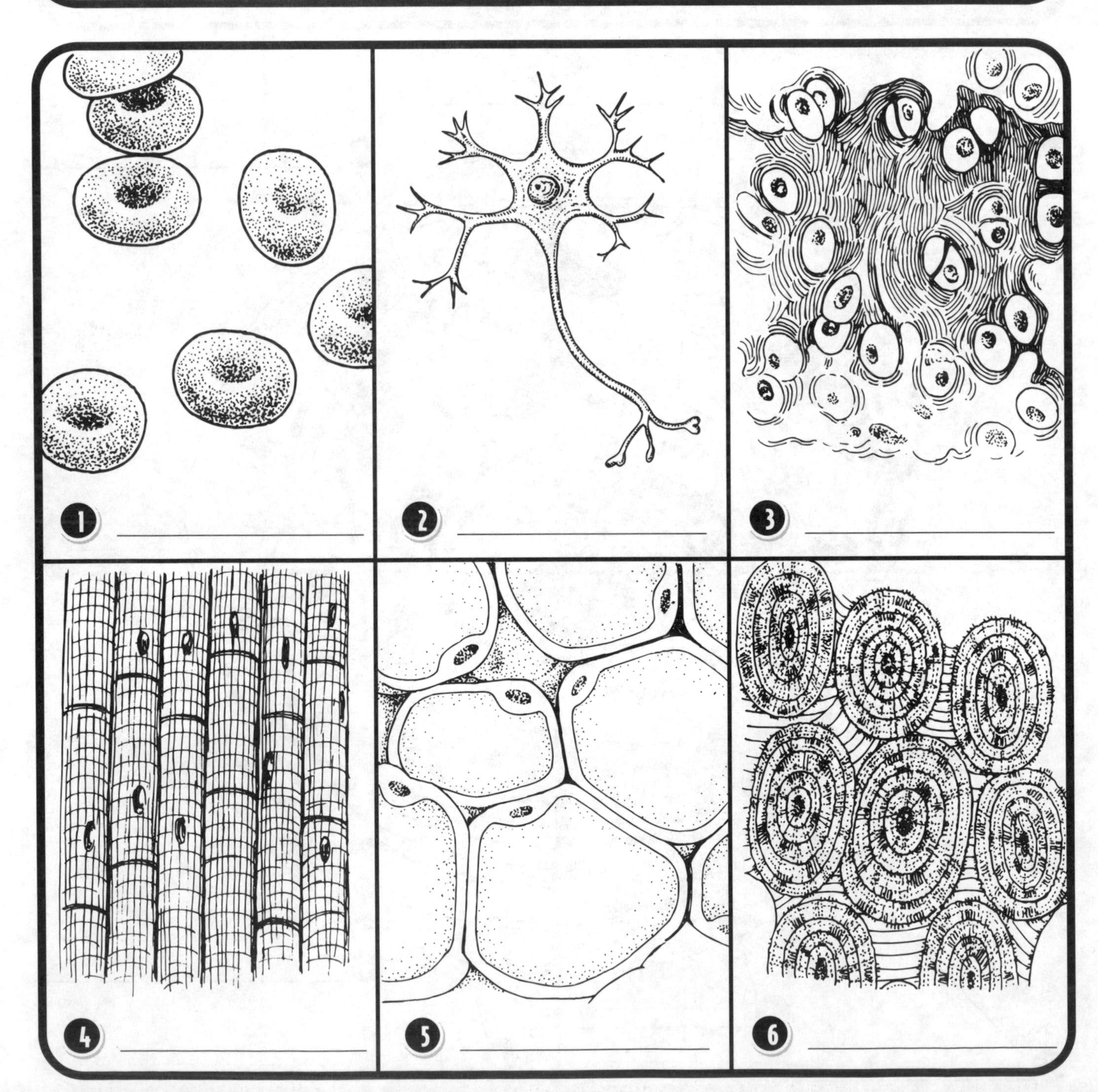

Name ______________________________ Date ______________

Types of Tissue

Many of the body's organs are made of a variety of tissues working together. There are four types of tissue: connective, epithelial, muscle, and nerve. Each type has a specialized function. Match the terms in the word box to the diagrams and descriptions below.

connective tissue	epithelial tissue	muscle tissue	nerve tissue
nerve fiber	collagen	nucleus	cell

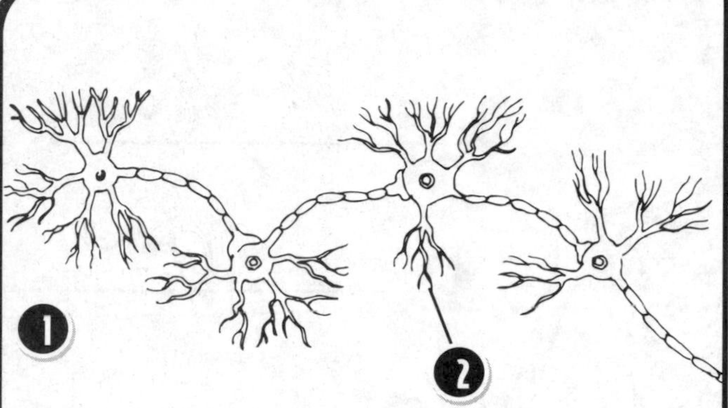

This tissue carries electrical impulses between the brain and body parts.

 ______________________________

2 ______________________________

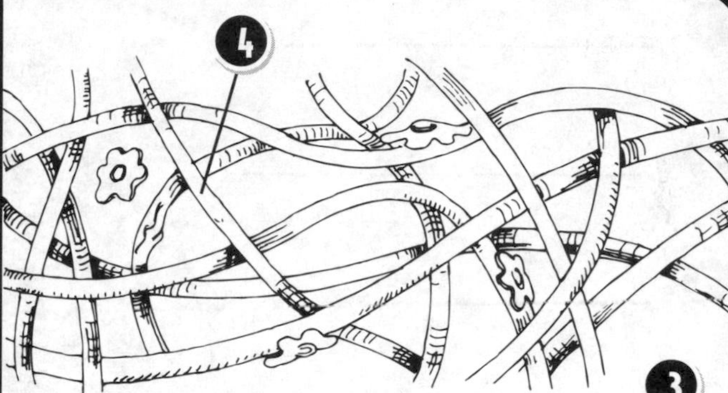

This tissue is made up of relatively few cells. It supports and connects other tissues.

3 ______________________________

 ______________________________

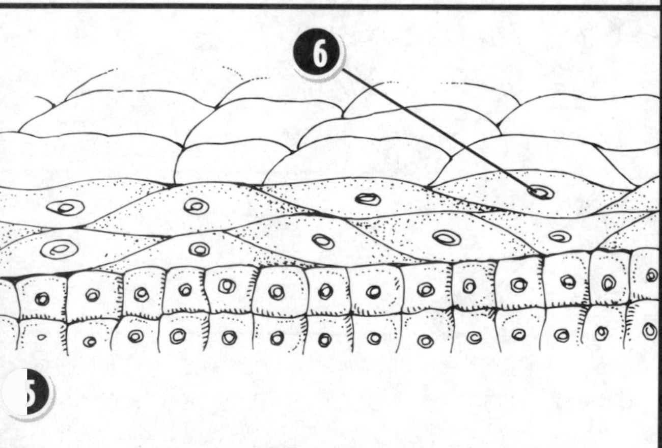

5

This tissue consists of tightly packed cells that cover the skin and line the hollow internal organs.

 ______________________________

 ______________________________

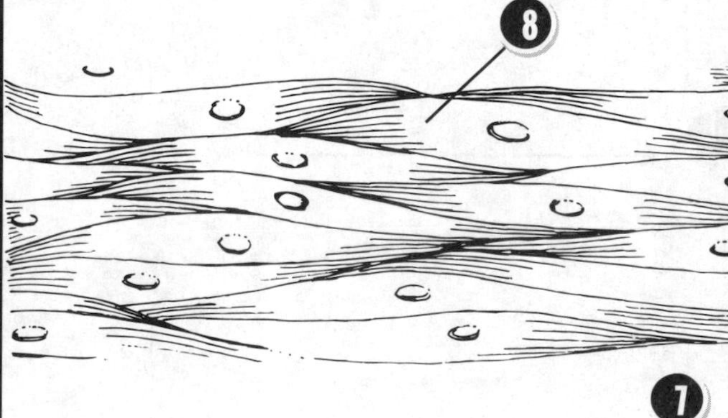

7

This tissue is made up of cells that can contract and relax.

7 ______________________________

8 ______________________________

Name ______________________ Date ____________

The Skeletal System

The skeletal system acts as our body's framework. The human adult skeleton has a total of 206 bones. Use the terms from the word box to label the major bones of the body.

cranium	coccyx	patella	pelvis	femur	clavicle
scapula	radius	mandible	rib cage	vertebrae	tibia

1 ______________________

2 ______________________

3 ______________________

4 ______________________

5 ______________________

6 ______________________

7 ______________________

8 ______________________

9 ______________________

10 ______________________

11 ______________________

12 ______________________

Name ______________________ Date ______________

Common Names for Fancy Words

Many bones of the skeletal system are known by more common terms. Match each term in the word box with its more commonly known description. Then match the number to each corresponding part on the diagram.

skull	tailbone	kneecap	hip bones	jawbone	backbone
shinbone	collarbone	thighbone	breastbone	shoulder blade	

1. ______________________ cranium
2. ______________________ mandible
3. ______________________ vertebrae
4. ______________________ tibia
5. ______________________ patella
6. ______________________ pelvis
7. ______________________ scapula
8. ______________________ clavicle
9. ______________________ coccyx
10. ______________________ sternum
11. ______________________ femur

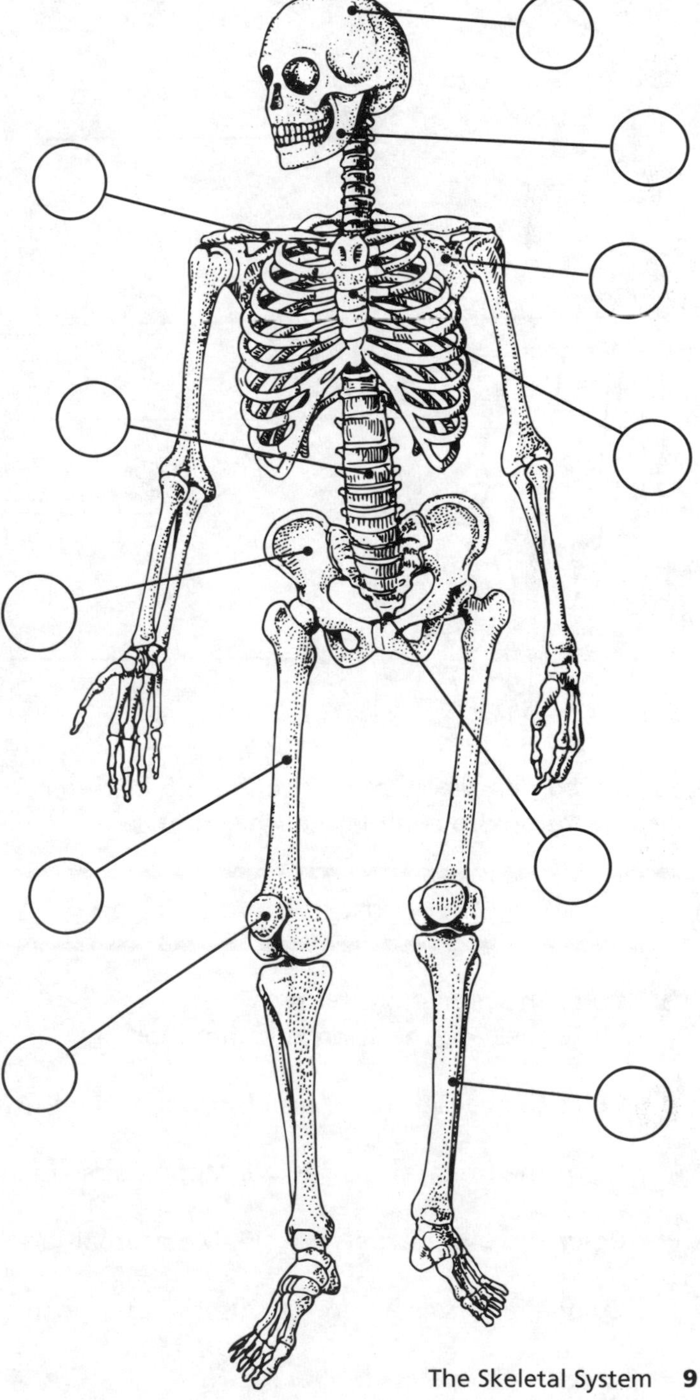

Name ______________________ Date ____________

Bones of the Hands and Feet

The human hand contains 27 bones divided into three groups. The human foot contains 26 bones divided into three groups. Use the terms in the word box to label the bones of the hands and feet.

phalanges	carpals	metacarpals	metatarsals	tarsals

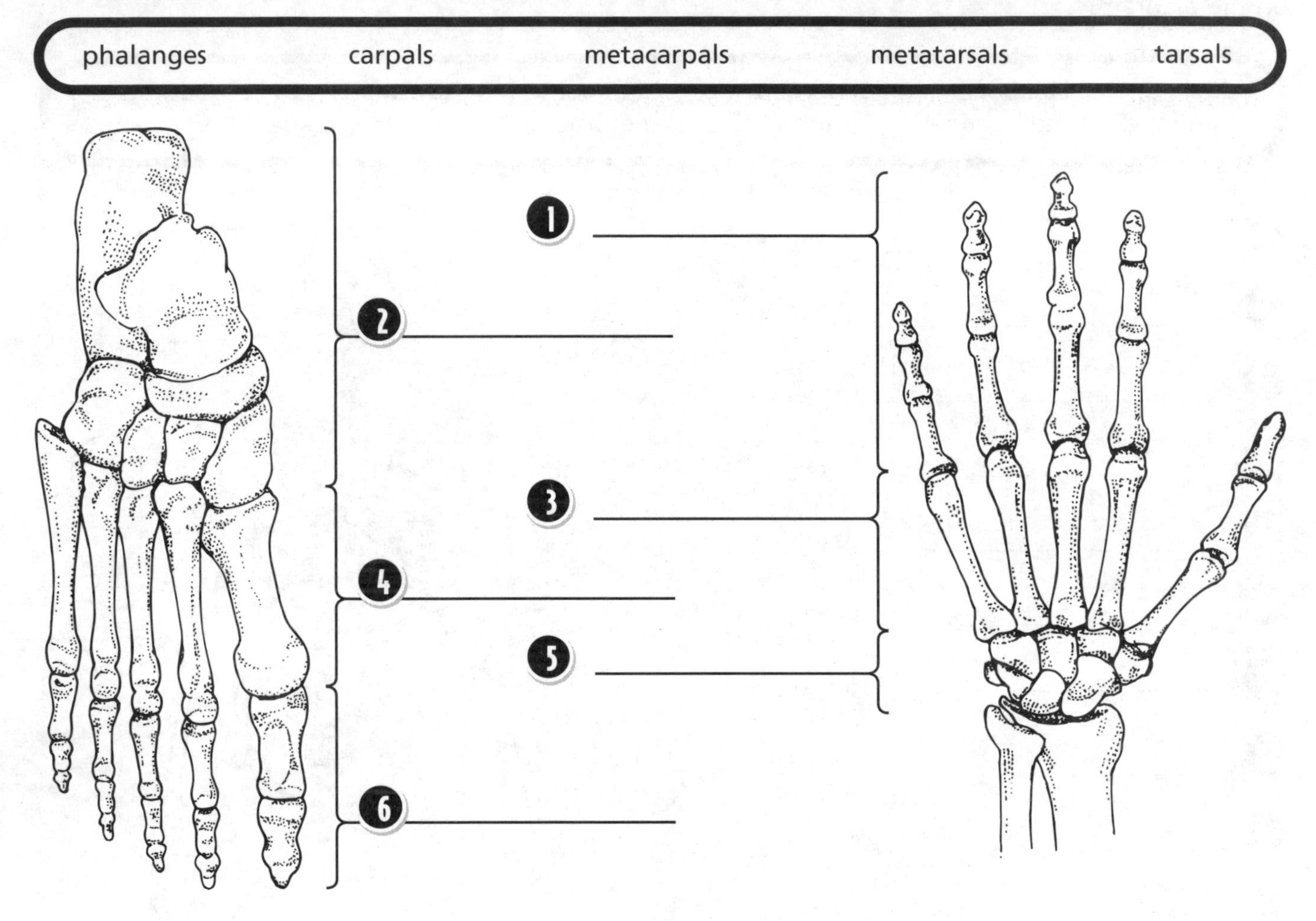

Use colored pencils or crayons to identify each section of the hand and foot. Use the common names in the word box to complete each sentence.

fingers	wrist	instep	toes	palm	ankle

7. Color the phalanges blue.
   Phalanges are the scientific name for ________________ and ________________.
8. Color the carpels red. Carpels are the scientific name for the ________________.
9. Color the metatarsals yellow. Metatarsals are also known as the ________________.
10. Color the metacarpals purple. Metacarpals are also known as the ________________.
11. Color the tarsals green. Tarsals is another name for the ________________.

Name ______________________ Date ____________

Bones of the Leg and Arm

The bones of the leg are designed for supporting the body and locomotion. The bones of the arm allow rotation and movement. The strongest bone in the body is the femur, located in the leg. Both the arm and the leg have a single bone in the upper portion and a pair of bones in the lower portion. Use the terms in the word box to label the diagram.

tibia	fibula	femur	patella	ulna	humerus	radius

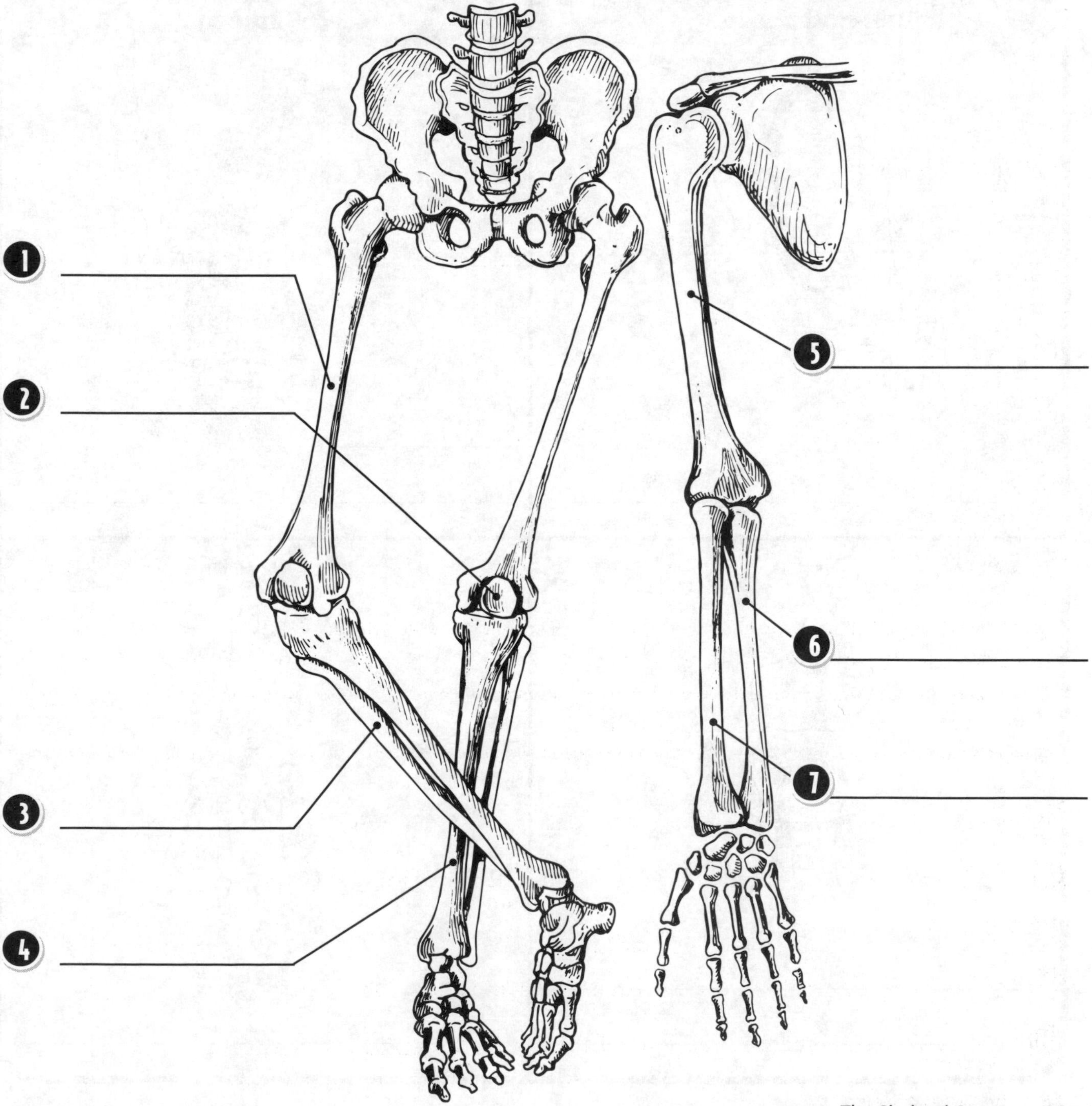

1 ____________

2 ____________

3 ____________

4 ____________

5 ____________

6 ____________

7 ____________

Name ______________________ Date ____________

Making Connections

The place where two or more bones meet is called a joint. Joints are either movable or immovable. There are four kinds of movable joints. Categorize each movable joint by listing it in the correct column.

shoulder	wrist	elbow	hip	knee	ankle
toe	neck	forearm	finger	backbone	

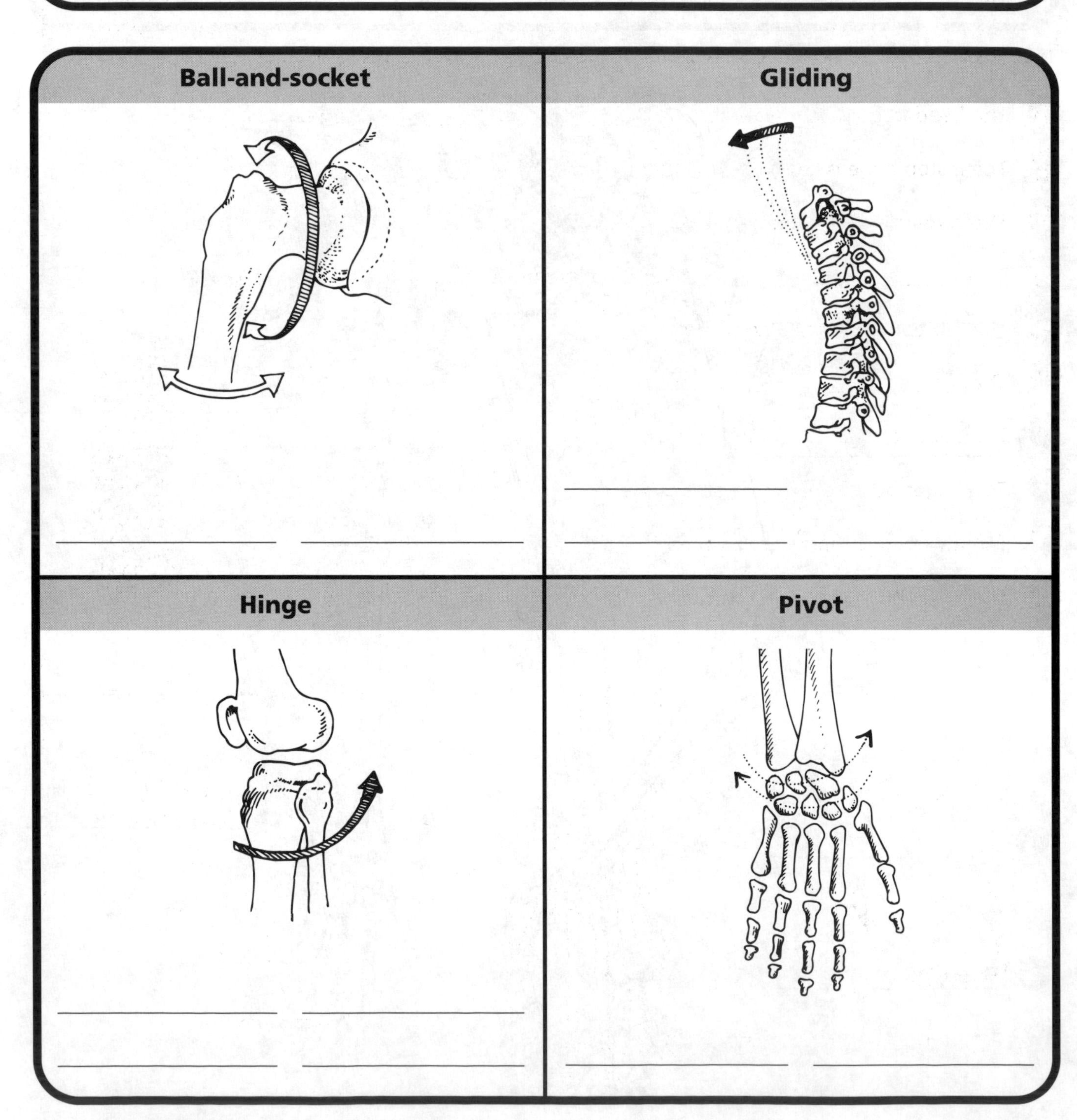

Name ______________________________ Date ______________

This Bone Is Connected to That Bone

One purpose of the skeletal system is to provide support for the body. In order to do that, the bones are interconnected through joints and cartilage to form a framework that supports and protects the limbs, torso, head, and organs. Use the terms in the word box to complete the sentences that describe how the bones are connected. Each term is used only once.

neck	shinbone	wrist	shoulder blade	backbone
toe	palm	head	instep	thighbone

1. The head bone is connected to the ______________________ bone.
2. The instep bone is connected to the ______________________ bones.
3. The forearm bone is connected to the ______________________ bone.
4. The ______________________ is connected to the knee bone.
5. The hip bone is connected to the ______________________.
6. The ______________________ is connected to the ankle bones.
7. The upper arm bone is connected to the ______________________.
8. The finger bones are connected to the ______________________ bones.
9. The jawbone is connected to the ______________________ bone.
10. The ankle bone is connected to the ______________________ bones.

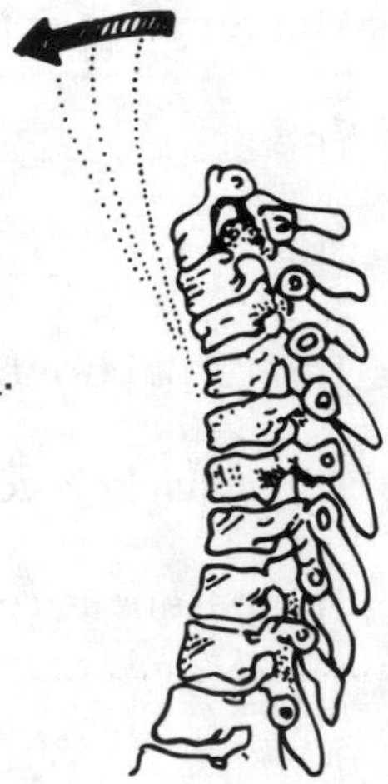

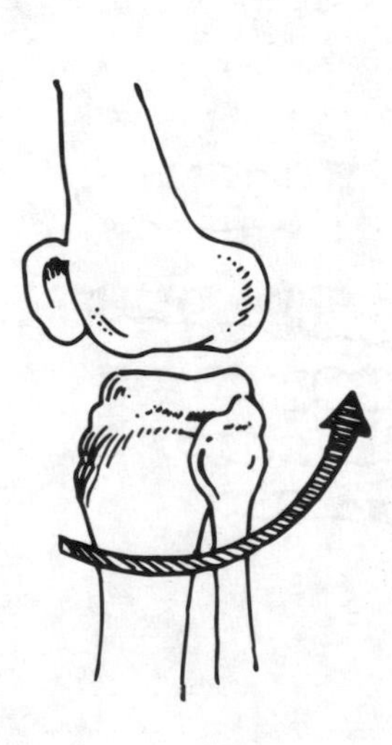

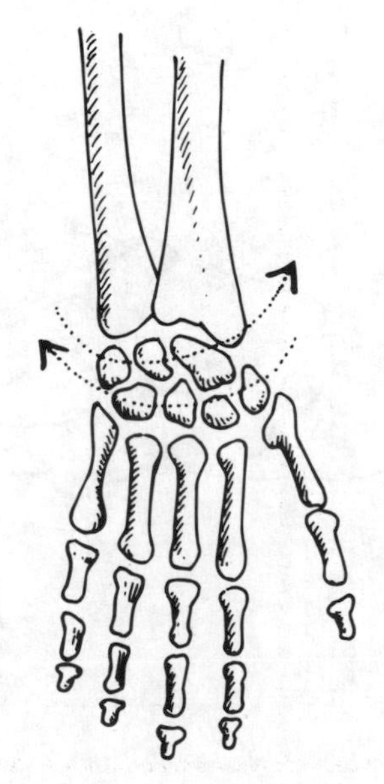

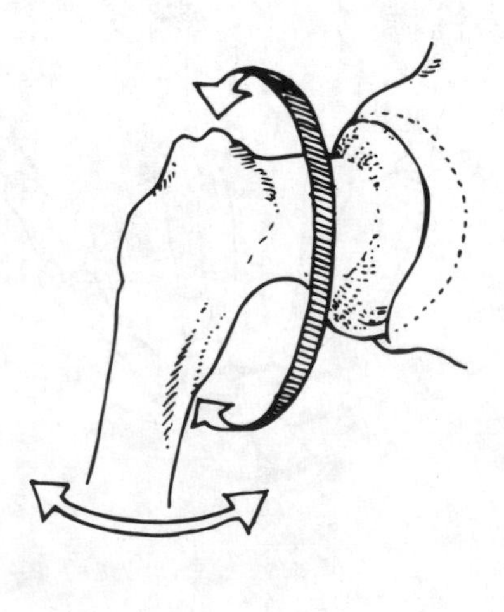

Name ______________________________ Date ______________

Scientifically Speaking

How well do you know the scientific names? Use the scientific terms in the word box to complete the sentences that describe how the bones are connected. Each term is used only once.

ulna	phalanges	tarsals	tibia	patella
humerus	vertebrae	cervical	sternum	femur

1. The cranium is connected to the ______________ bones.
2. The metatarsals are connected to the ______________.
3. The tibia is connected to the ______________.
4. The ______________ is connected to the pelvic bone.
5. The femur is connected to the ______________.
6. The ______________ is connected to the rib bones.
7. The ______________ are connected to the coccyx.
8. The carpal bones are connected to the radius and ______________.
9. The clavicle is connected to the ______________.
10. The patella is connected to the ______________ and fibula.

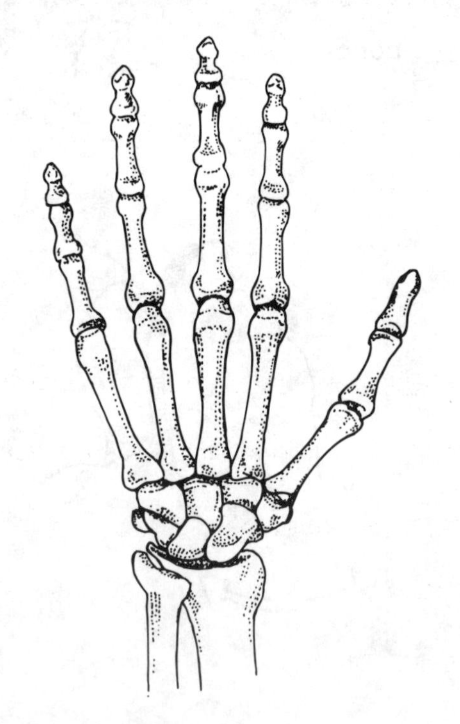

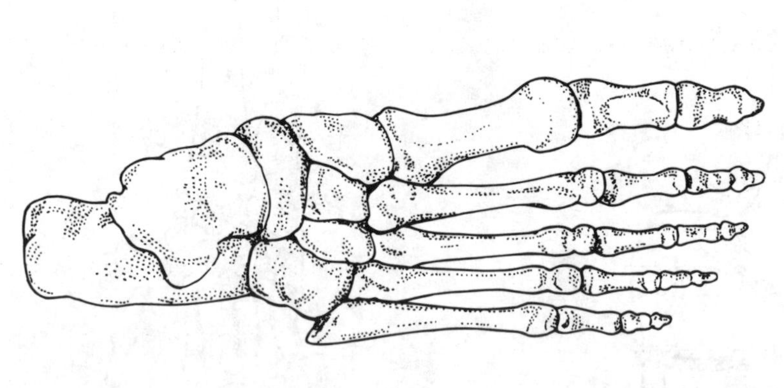

Name ______________________________ Date ______________

Parts and Function of a Bone

Bones are complex organs that function in movement, protection, support, mineral storage, and blood cell formation. Human bones are as tiny as the stirrup in the ear or as large as the femur in the leg. All have a similar structure. Use the terms in the word box to label the cross section of a femur.

marrow compact bone cartilage periosteum spongy bone blood vessels

1. ______________________________
2. ______________________________
3. ______________________________
4. ______________________________
5. ______________________________
6. ______________________________

Use the terms from the word box above to complete each sentence.

7. The ______________ produces new red blood cells for the body.
8. The ______________ is a covering of the bone to which ligaments and tendons are attached.
9. ______________ is inner tissue that provides strength to the bone without a lot of added weight.
10. ______________ carry nutrients to and wastes away from the bone cells.
11. ______________ is connective tissue between bone joints.
12. ______________ is dense outer tissue that resists shock and stress.

Name ______________________ Date ____________

Your Teeth

Teeth are hard bony structures in the mouths of humans and animals. The adult human has 32 teeth, 16 in the upper jaw and 16 in the lower jaw. There are four types of teeth. Use the terms in the word box to label the diagrams.

neck	cementum	root	enamel	crown	pulp	dentin
root canal	nerve	incisors	canines	bicuspids	molars	

Parts of a Tooth

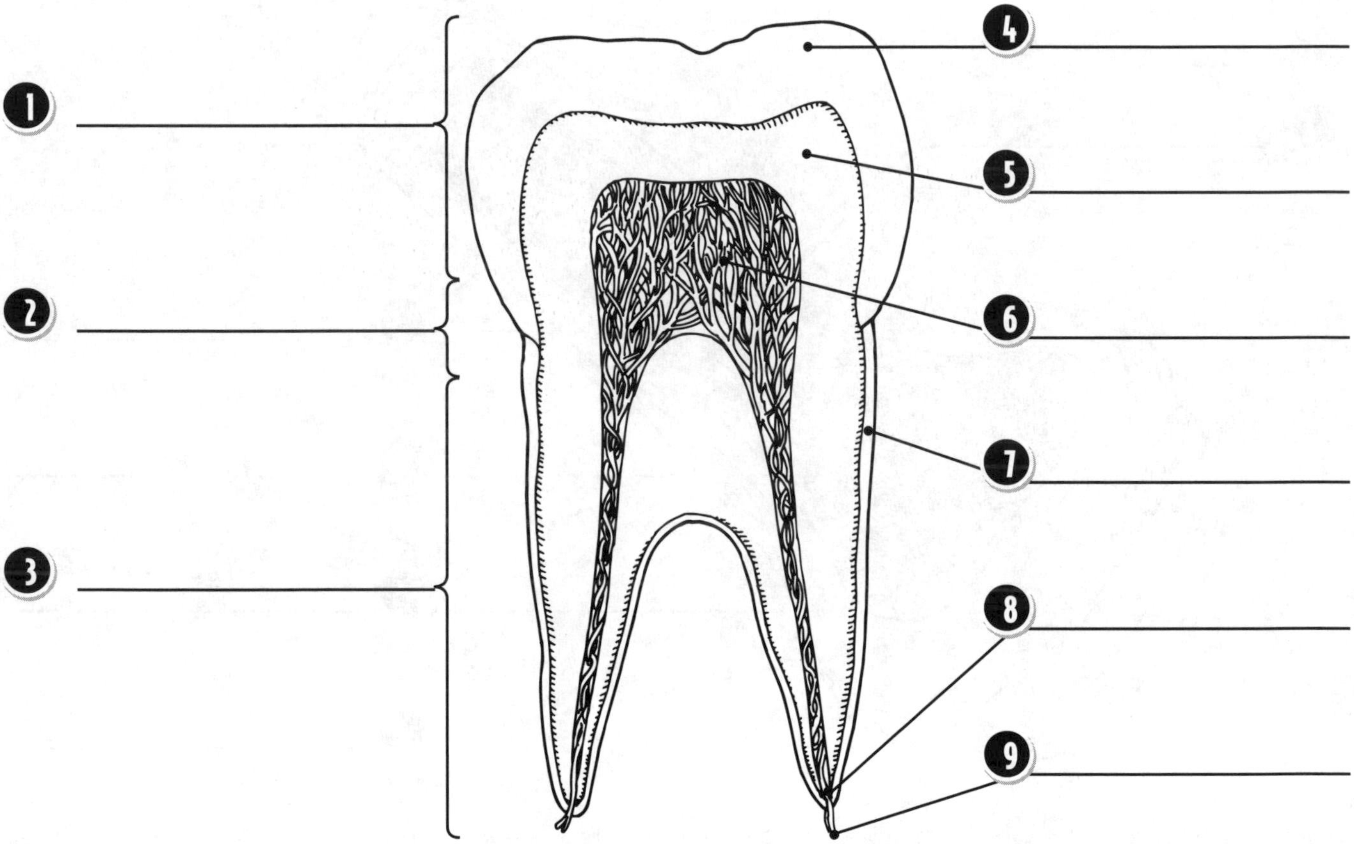

1. ______
2. ______
3. ______
4. ______
5. ______
6. ______
7. ______
8. ______
9. ______

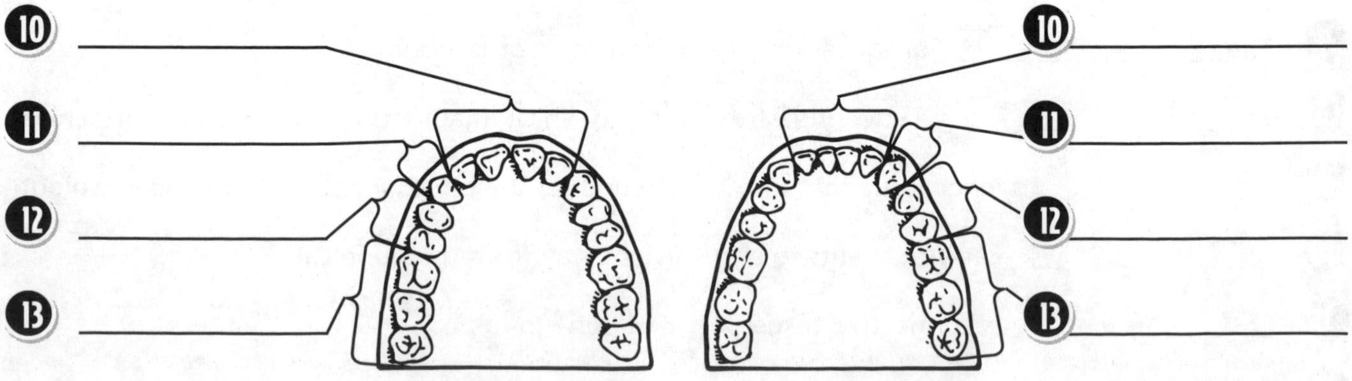

10. ______ 10. ______
11. ______ 11. ______
12. ______ 12. ______
13. ______ 13. ______

Upper Teeth **Lower Teeth**

Name ______________________ Date ____________

Bones of the Head

The skull is the bony structure located at the top of the spinal column in all vertebrate animals. The skull encases and protects the brain. In human infants, the skull contains areas of cartilage, becoming harder bone as a child matures. The human skull is divided into two sections—the cranial and facial bones. Use the terms in the word box to label the diagrams.

mandible	maxillary nasal bone	frontal bone	parietal bone
cranium	temporal bone	occipital bone	teeth
eye socket	vertebrae	nose cartilage	

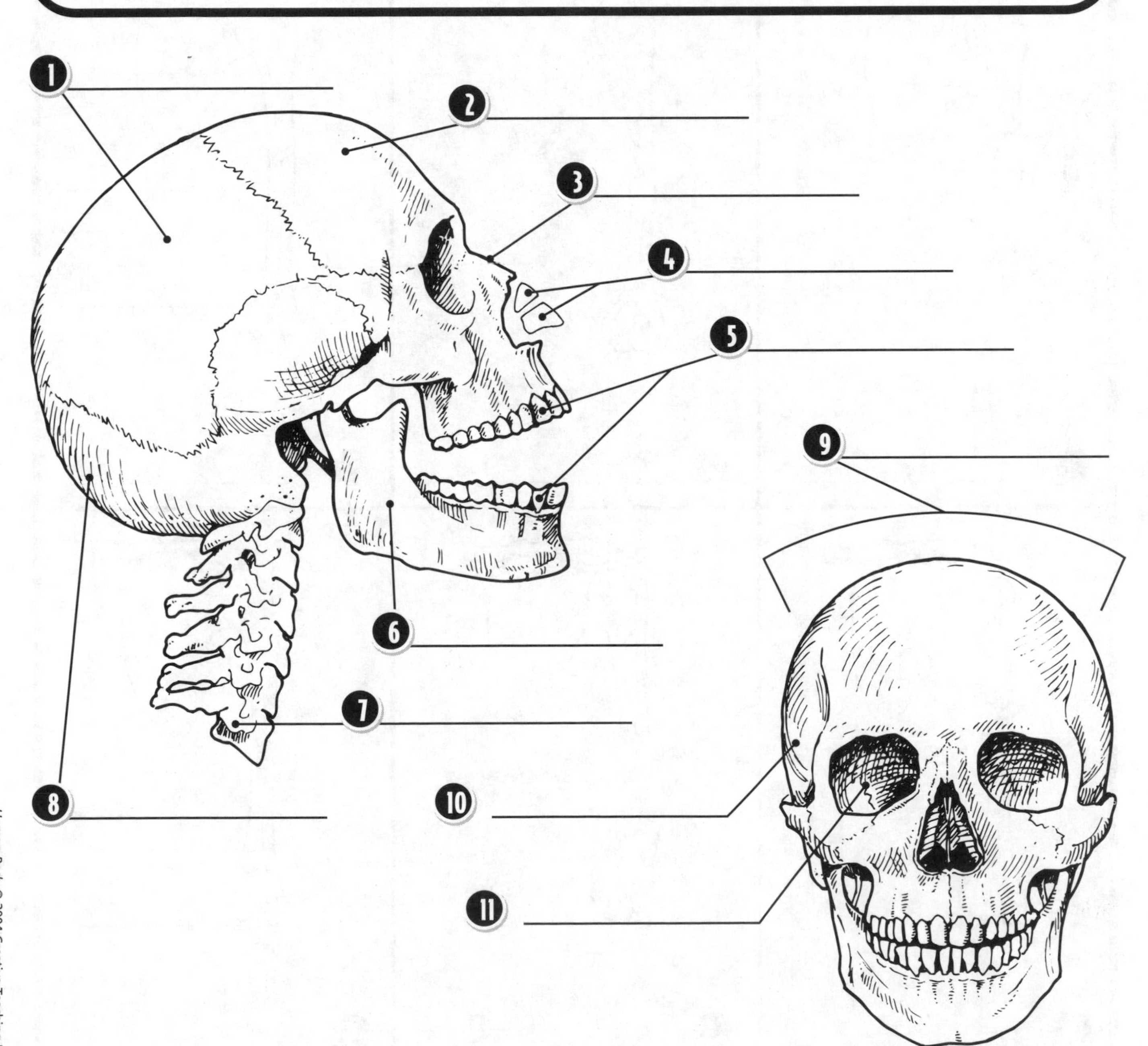

Name ______________________ Date ____________

Injuries to Bones

A fracture is a break in a bone. Each type of fracture has a different name based on how the bone is broken. Use the terms in the word box to label each type of fracture.

closed	open	comminuted	multiple	spiral	greenstick

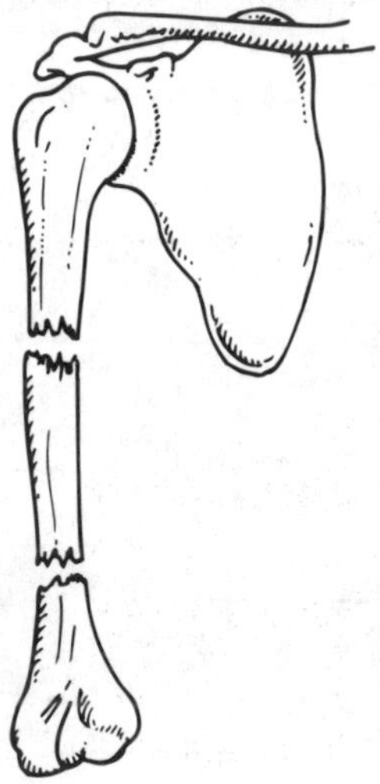

1. ______________
2. ______________ 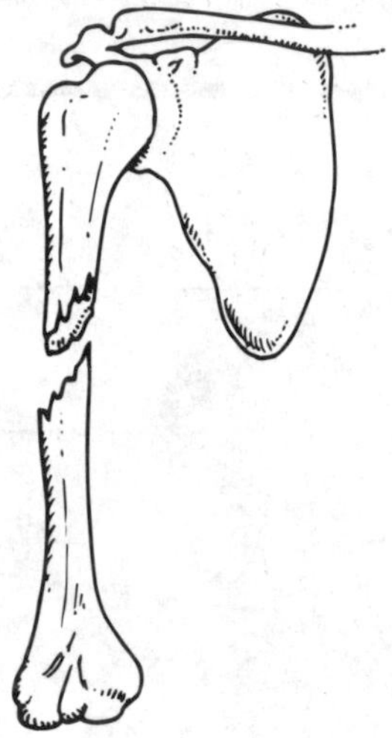
3. ______________ 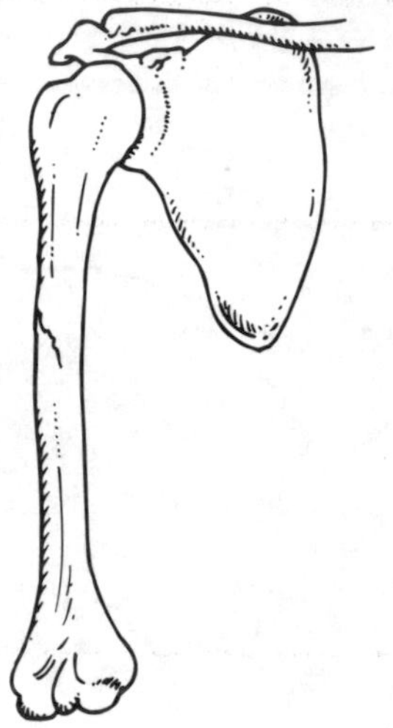
4. ______________ 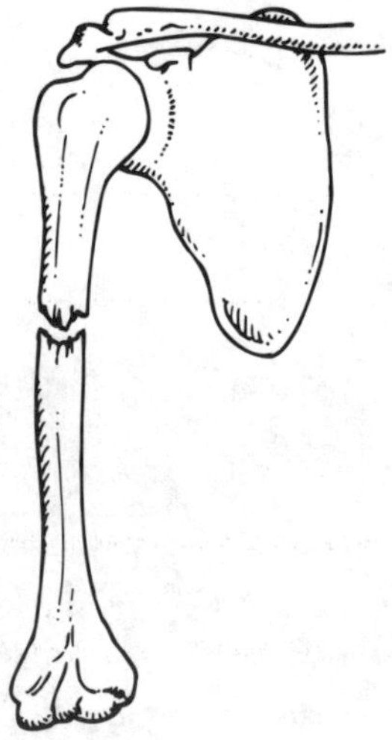
5. ______________ 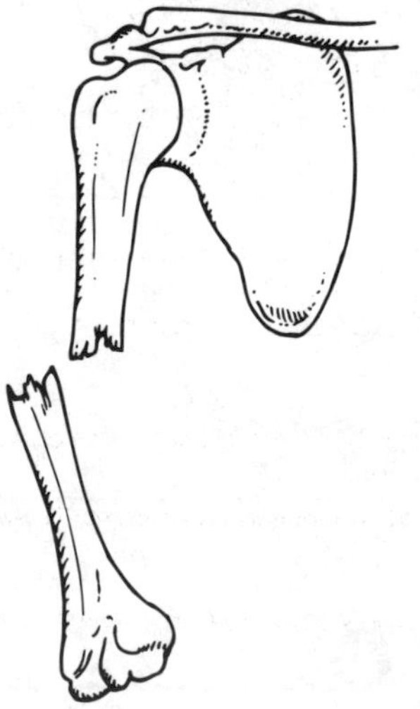
6. ______________

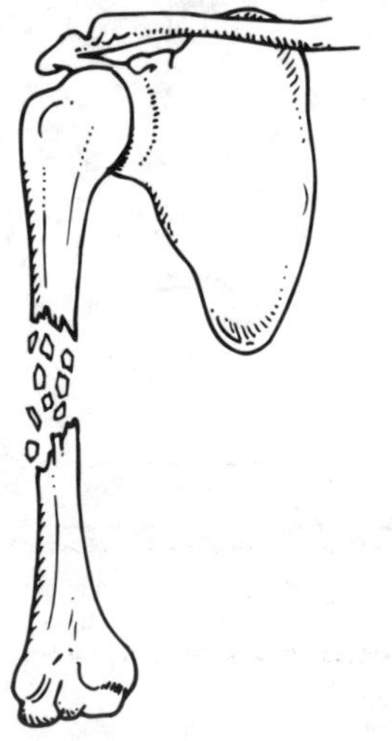

Name ______________________________ Date ______________

The Backbone

The spinal column is the structure of bone or cartilage surrounding and protecting the spinal cord in humans. It is also called a vertebral column, spine, or backbone. Use the terms in the word box to complete the diagrams.

vertebra	disc	cervical region	coccygeal region
thoracic region	lumbar region	sacrum	

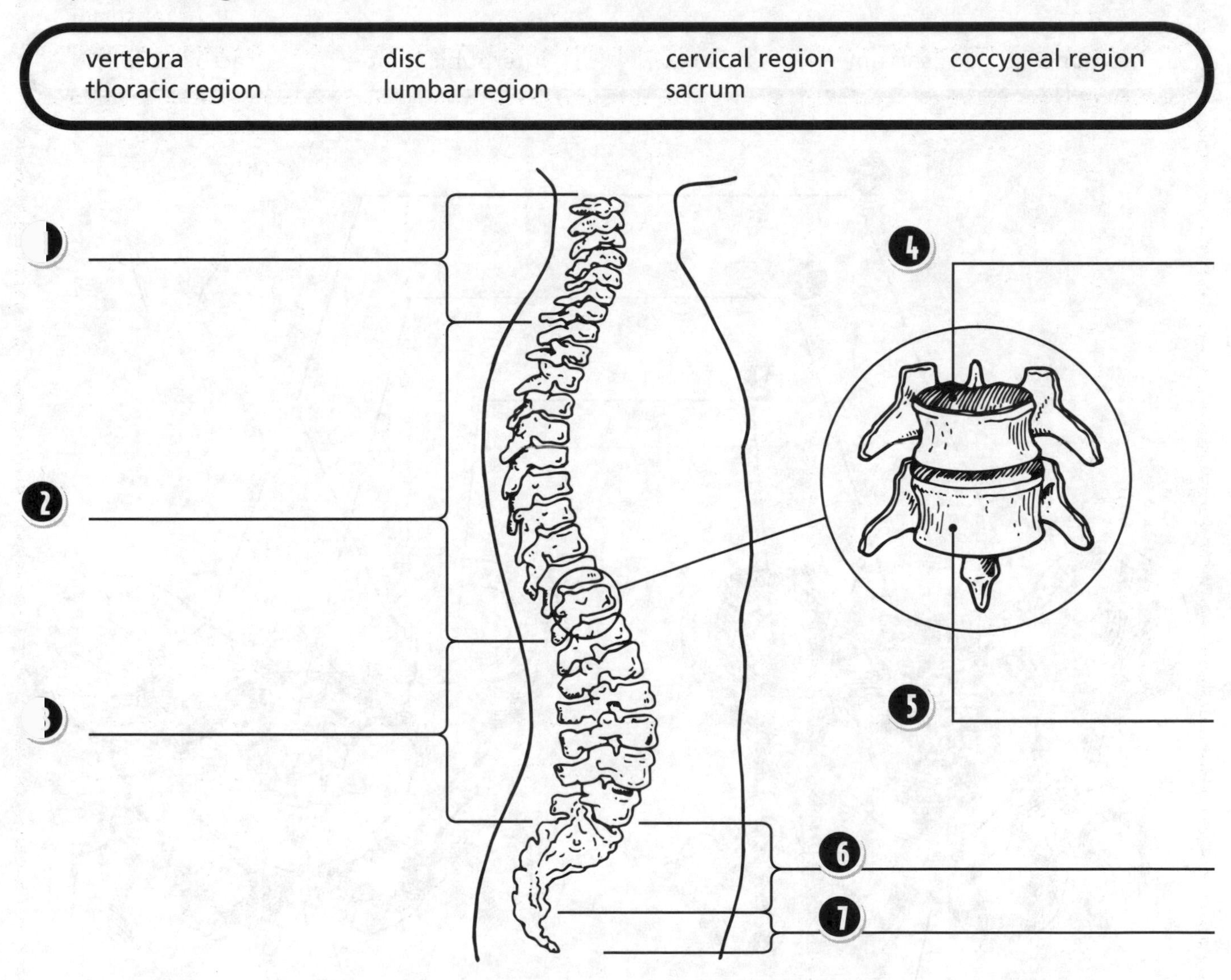

Use the terms in the word box to match the common name to its scientific name.

tailbone	neck	chest	lower back	pelvic girdle

8 ______________ cervical region
9 ______________ coccygeal region
10 ______________ thoracic region
11 ______________ lumbar region
12 ______________ sacral region

Name ______________________ Date ____________

The Human Pelvis

The framework of bones that supports the lower part of the abdomen is called the pelvis. The male pelvis is heart-shaped and narrow. The female pelvis is wider and flatter, allowing for childbirth. Use the terms in the word box to complete the diagrams.

hip bone	sacrum	coccyx	interpubic joint	sacroiliac joint

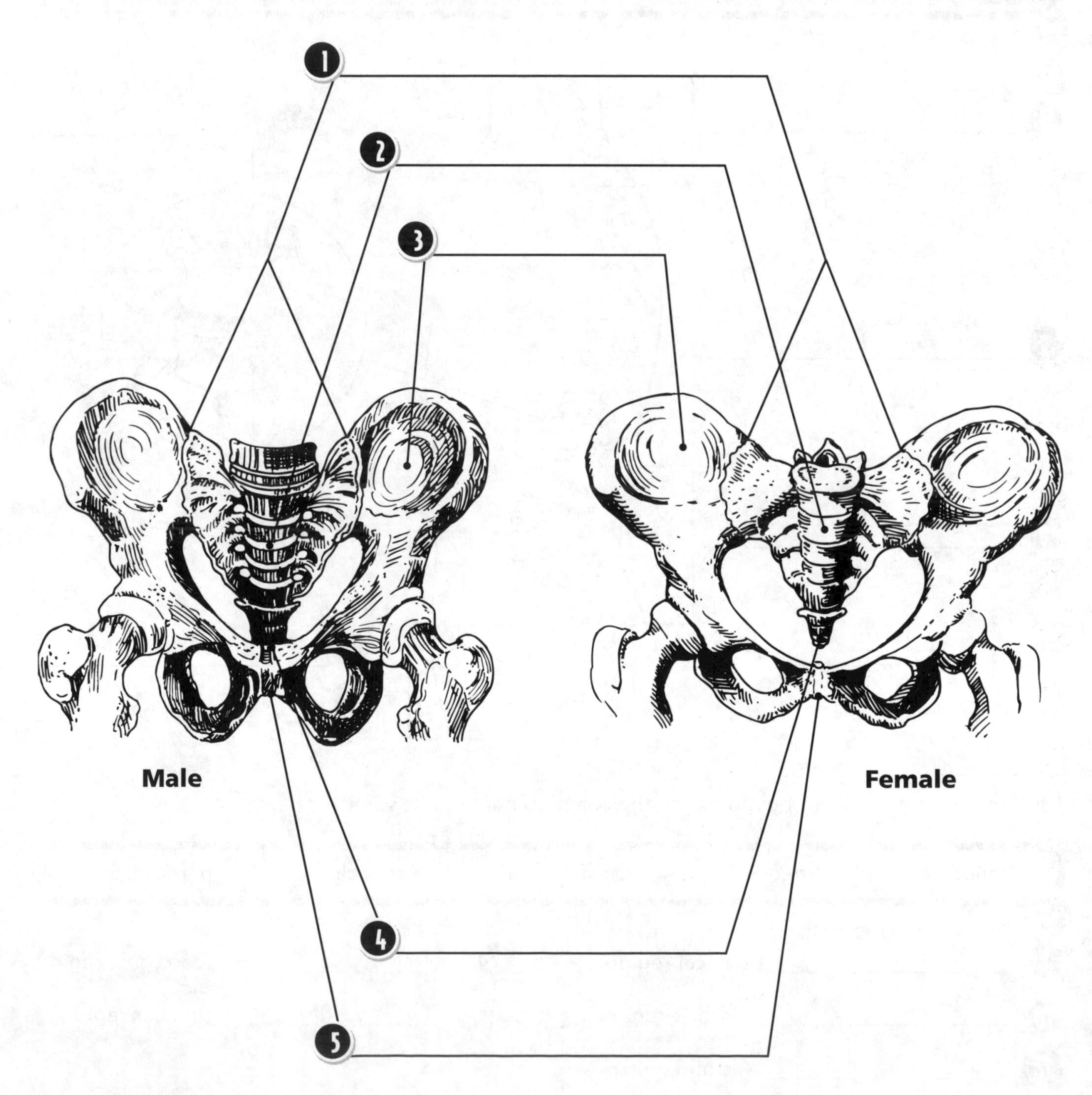

Name ______________________ Date ____________

What Do You Know About Bones?

Review the skeletal system. Match each term in the word box to its definition.

metatarsals	mandible	periosteum	ball-and-socket	bicuspid
patella	thoracic region	dentin	ulna	scapula
clavicle	femur			

1. ______________ This middle section of five bones between the toes and the ankle forms the arch of your foot.
2. ______________ This is a flat, triangular bone located to the front of your knee joint.
3. ______________ This is the main part of a tooth, between the enamel and the pulp.
4. ______________ This part of the spinal column coincides with your chest.
5. ______________ This is the outer covering of a bone to which tendons and muscles are attached.
6. ______________ Your shoulder is this type of joint.
7. ______________ This hinged bone allows you to chew your food.
8. ______________ This is the largest and strongest bone in the human body.
9. ______________ A type of tooth, this has two points or cusps.
10. ______________ This large, flat, triangular bone forms the back part of the shoulder.
11. ______________ This bone extends from the sternum to the scapula.
12. ______________ This is a bone located in the forearm.

Name ____________________ Date __________

The Muscular System

There are hundreds of muscle groups in the human body. Some groups can be seen under the surface of the skin. Others are deep within the body. Muscles have the ability to contract, usually in response to messages from the nervous system. Use the terms in the word box to label these muscle groups.

pectorals	quadriceps	deltoids	gastrocnemius
biceps	sternocleidomastoids	intercostals	triceps

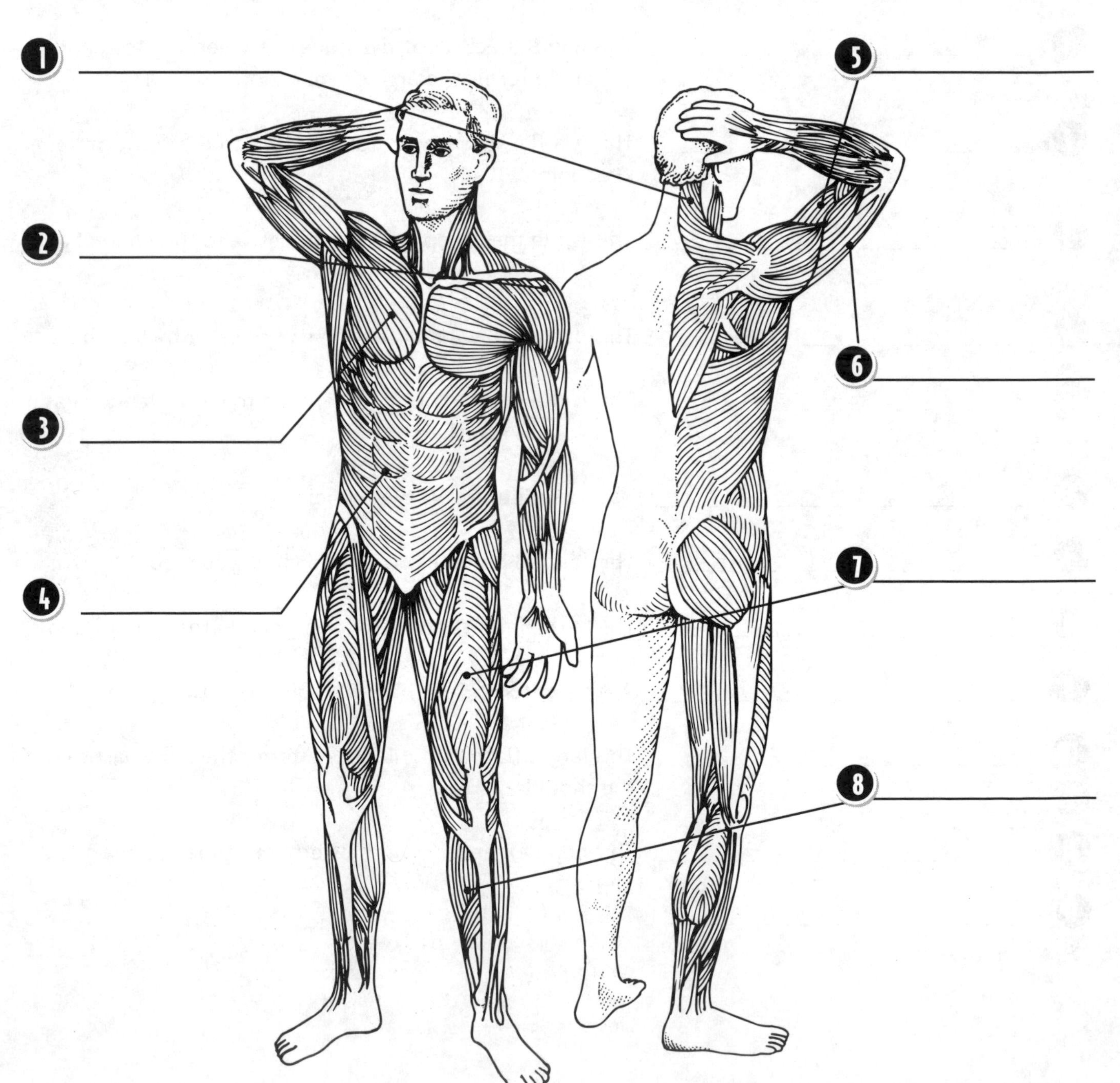

Name ______________________________ Date ______________

Work Those Muscles!

The muscles of the body control movement. While the bones provide support, the contracting and relaxing of muscles allows us to move. Match each term in the word box to its scientific name or function.

biceps	shoulder muscles	thigh muscles	calf muscles	chest muscles
neck muscles	abdominals	triceps	gluteus maximus	trapezius

1. ______________________ Lift a glass from the table to your mouth and you are using these muscles.
2. ______________________ Also known as quadriceps, you'll feel the burn here if you climb a lot of stairs.
3. ______________________ Also known as your deltoids, it always feels so good to get a back rub here.
4. ______________________ You spend a lot of time sitting on these when you are at school.
5. ______________________ Tiptoe around a while. You'll feel it in these gastrocnemius muscles.
6. ______________________ Sit-ups keep these intercostals strong.
7. ______________________ Look up. Look down. Look side to side. These sternocleidomastoids have a lot of work to do.
8. ______________________ This large, three-headed muscle forms the back of the upper arm.
9. ______________________ This large flat muscle of the back allows you to raise your head and shoulders.
10. ______________________ How many push-ups can your pectorals handle? A lot, if these muscles are strong.

Name ______________________ Date ____________

Types of Muscle Tissue

There are three kinds of muscles found in the body. Use the terms in the word box to label each diagram. Then classify the job each type of muscle can do using the phrases in the word box.

skeletal muscles	smooth muscles	cardiac muscles
move food through the digestive tract	bend arms and legs	squeeze the bladder
contract blood vessels	close a fist	create a smile or a frown
maintain a heartbeat	keep blood pumping	found only in the heart

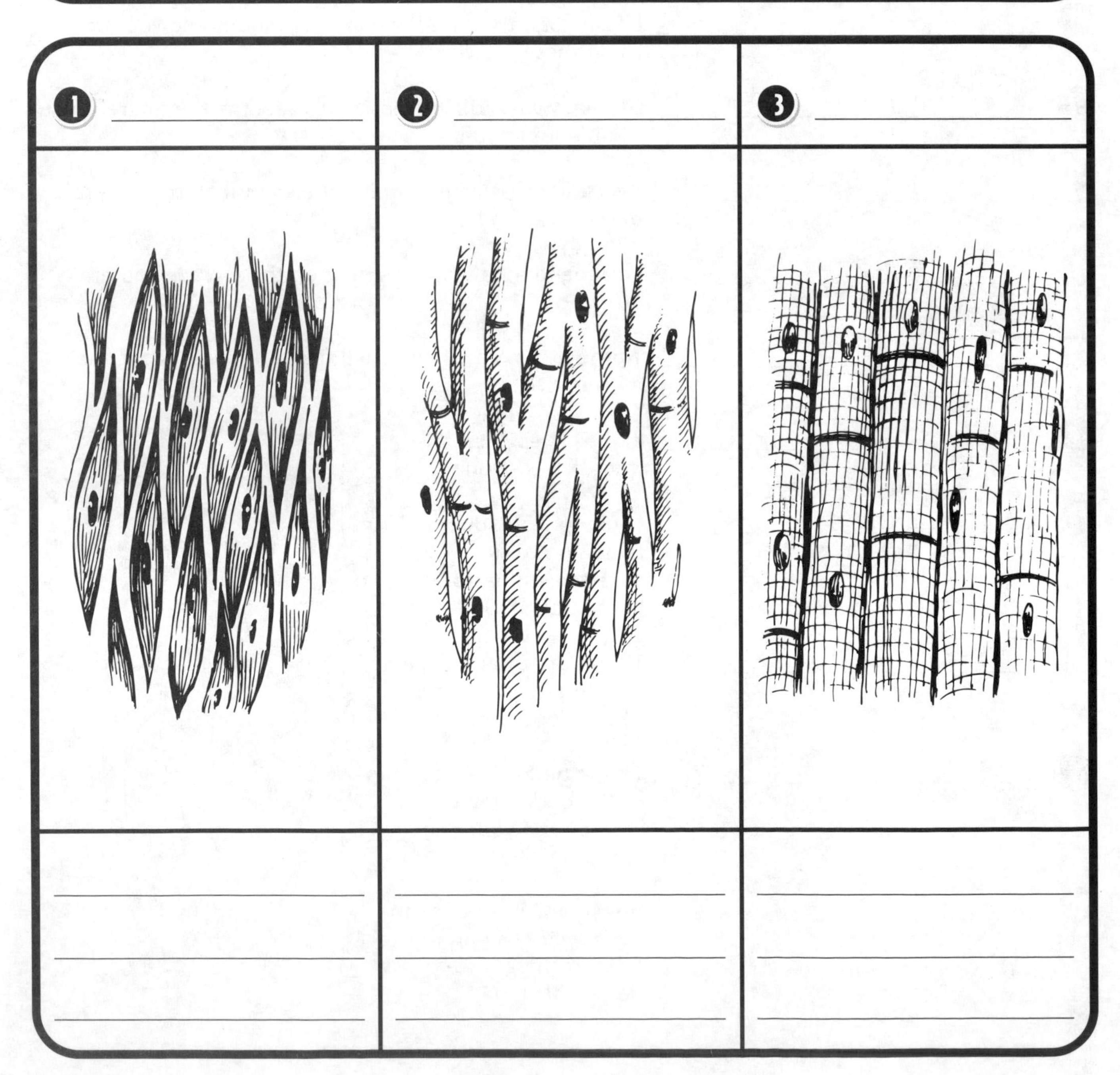

Name ______________________________ Date ______________

Muscular Connections

The skeletal muscles allow our bodies to move. These muscles are attached to bones. When they contract, the muscles get fatter and shorter. When they relax, the muscles get longer and thinner. Use the terms in the word box to label the diagrams.

biceps	triceps	tendon	biceps relaxed
triceps relaxed	quadriceps relaxed	hamstring relaxed	biceps contracted
triceps contracted	quadriceps contracted	hamstring contracted	

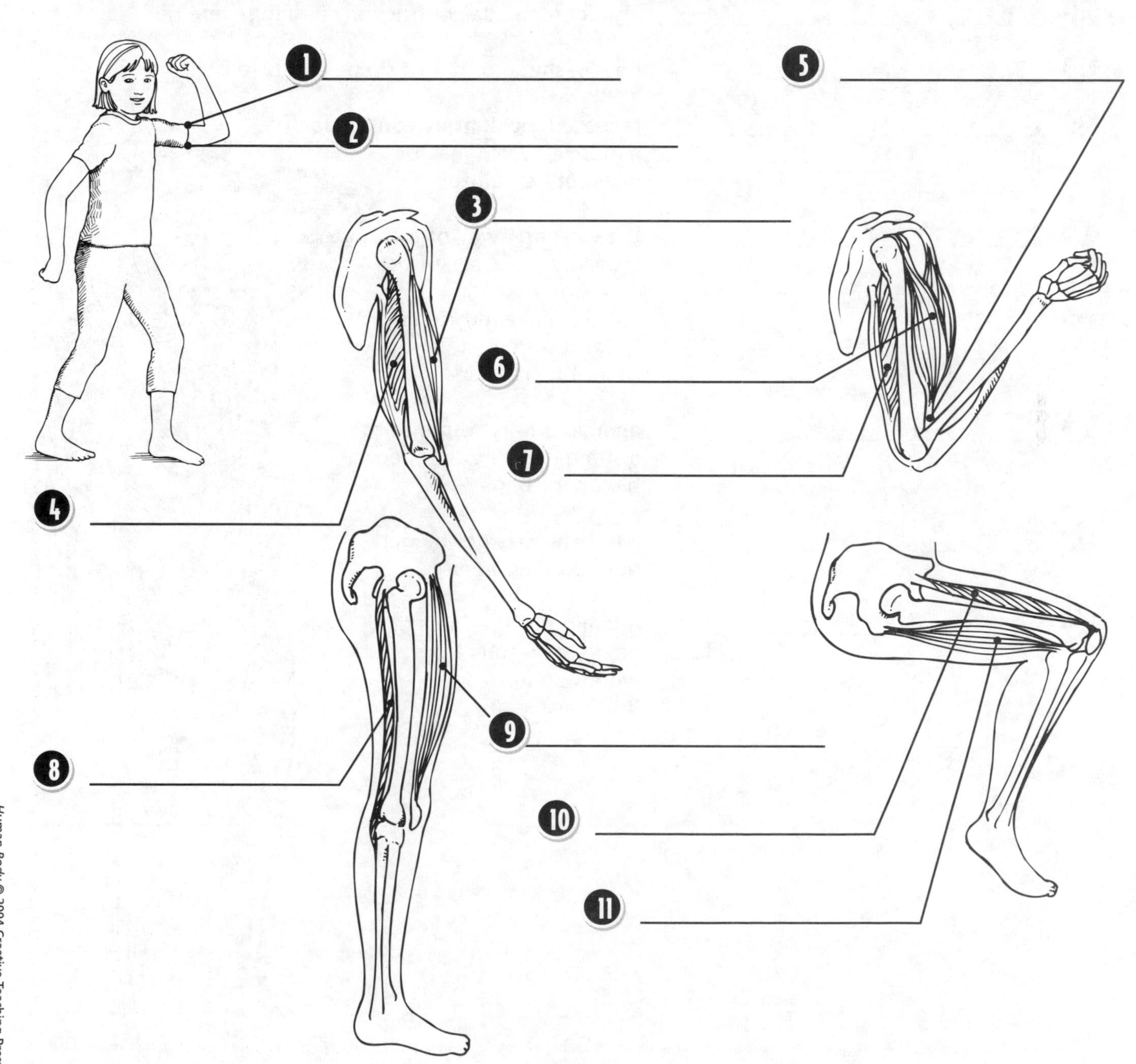

Name ______________________________ Date ______________

Muscular Actions

Muscles are meant to move and react to stimuli. Sometimes things happen when muscles do not work correctly, or when they are worked too much or too little. Illness can also affect muscles. Match each term in the word box to its description.

muscle cramp	muscle spasm	atrophy	hypertrophy
muscle ache	reflex	muscle tone	muscular dystrophy

1. ______________________ a sudden muscle contraction causing severe pain
2. ______________________ how in shape or out of shape a muscle is
3. ______________________ repeated involuntary contraction of a muscle, which may or may not be painful
4. ______________________ the wasting away of a muscle due to lack of use
5. ______________________ a progressive muscle disorder that causes muscle atrophy that cannot be reversed
6. ______________________ an involuntary response to a stimulus, such as a sneeze, hiccup, or blink
7. ______________________ when muscles get bigger due to excessive use
8. ______________________ a feeling of tiredness or pain that results from working muscles

Name ______________________ Date ____________

The Circulatory System

The circulatory system carries oxygen and nutrients to the cells and carries wastes away from the cells. Use the terms in the word box to label the diagram.

heart	liver	capillaries	lungs	artery	kidneys	vein

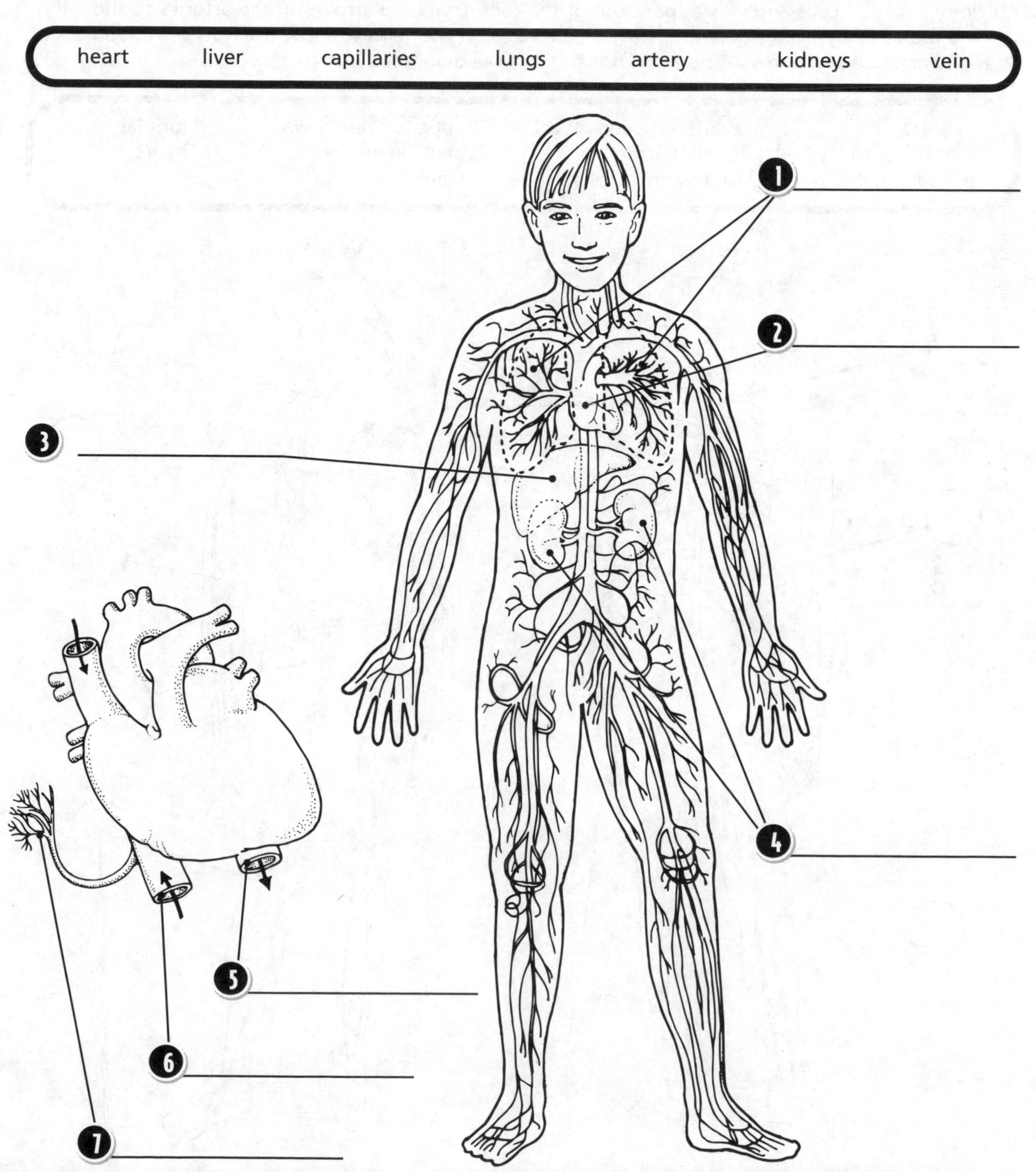

Name ______________________________ Date ______________

Veins and Arteries

Arteries carry blood away from the heart to the parts of the body. Veins carry blood back to the heart. Capillaries are very small, thin-walled blood vessels that allow the exchange of oxygen and nutrients into the cells and wastes back out of the cells. Draw red arrows on the arteries to show the flow of blood away from the heart. Draw blue arrows on the veins to show the flow of blood back to the heart. Use the terms in the word box to label the diagram.

1. aorta	2. inferior vena cava	3. superior vena cava	4. jugular vein
5. carotid artery	6. renal artery	7. renal vein	8. heart
9. capillaries	10. femoral artery	11. femoral vein	

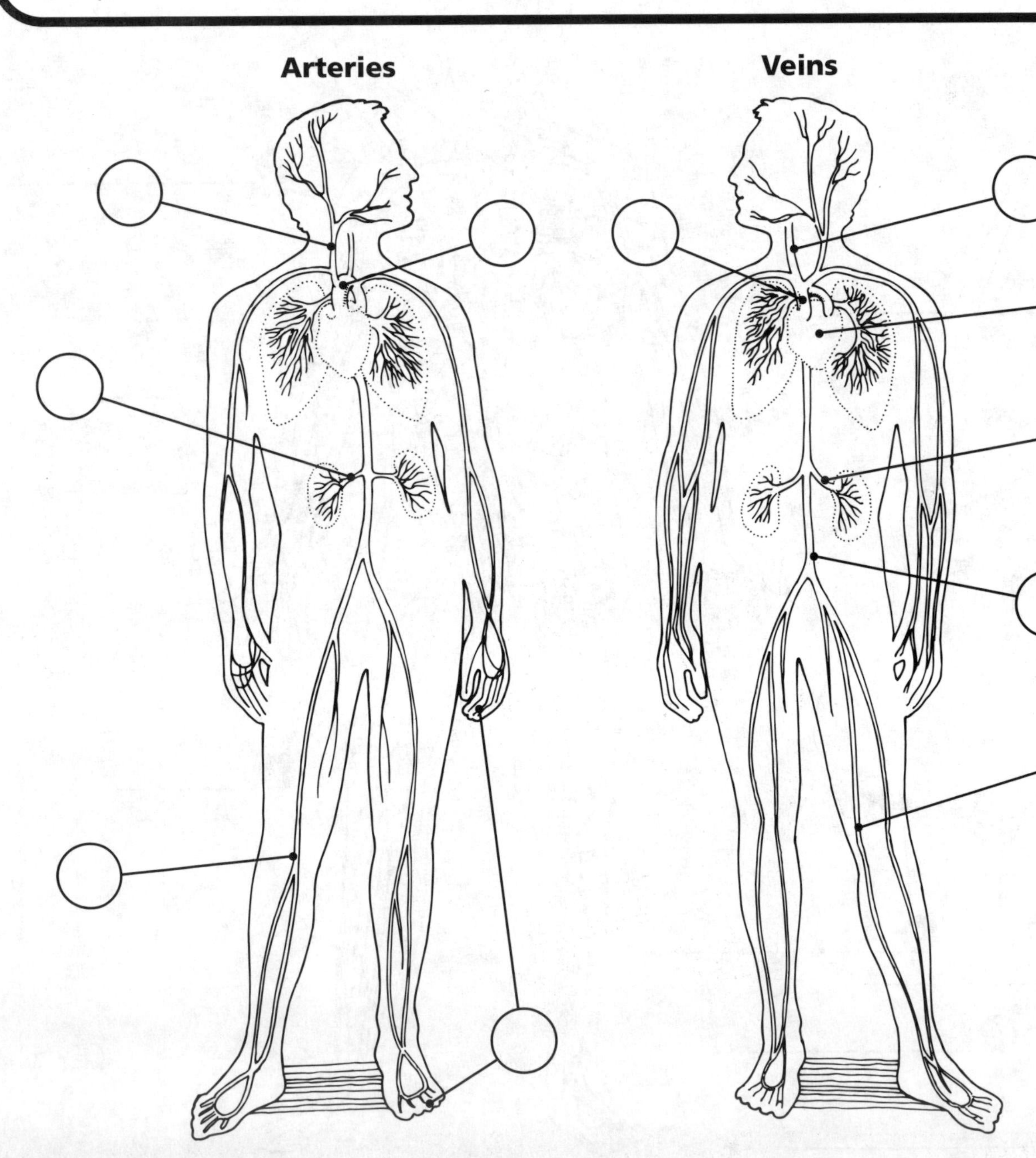

Name ______________________________ Date ______________

The Heart

The heart is a pump that beats continuously in order to deliver oxygen to the cells. It is shaped like an upside-down pear and is about the size of a closed fist. Use the terms in the word box to label the parts of the heart.

left atrium	left ventricle	right atrium	right ventricle
vena cava	pulmonary veins	pulmonary artery	aorta

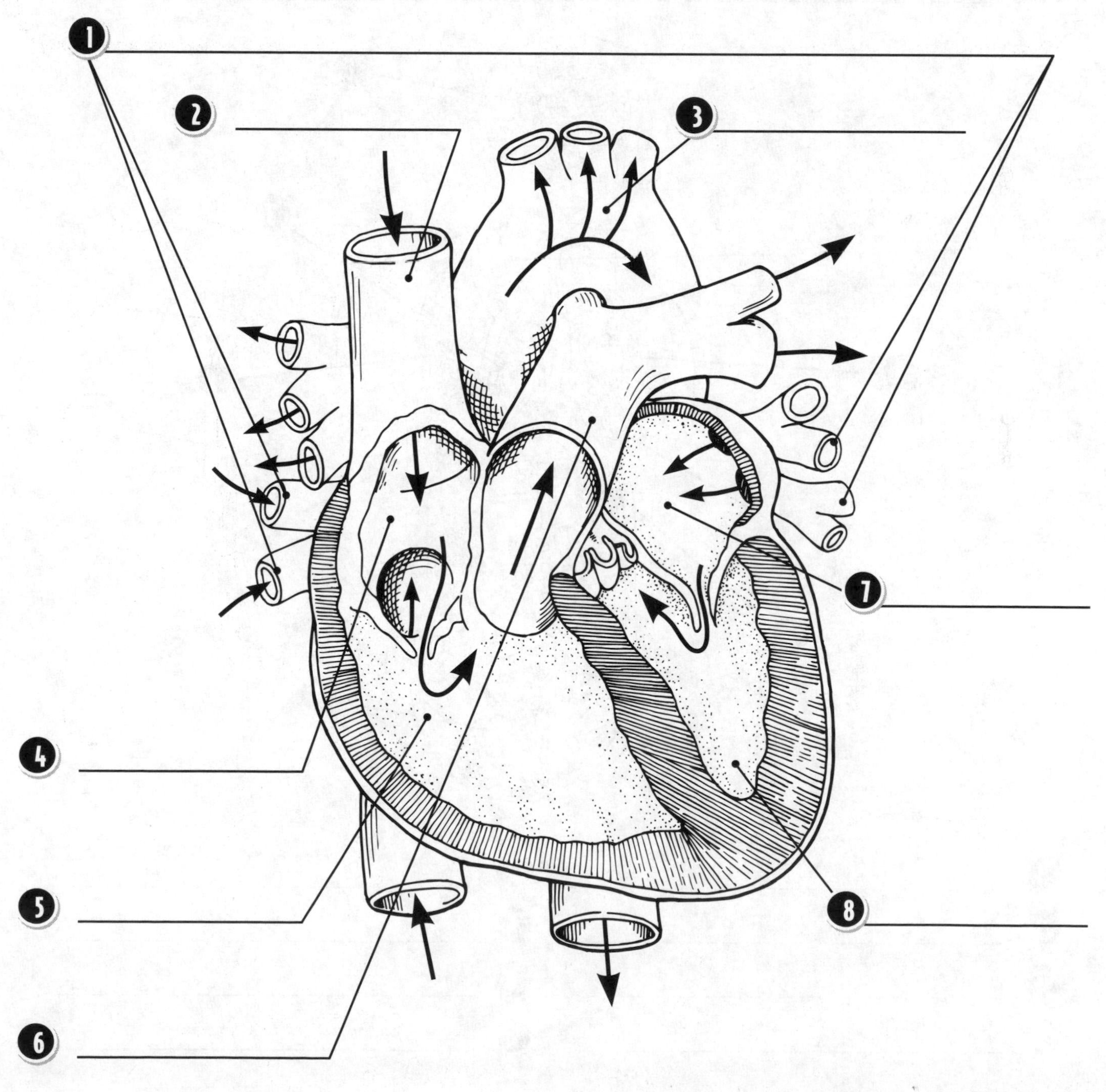

Name ______________________________ Date ______________

The Heart Has a Job to Do

The human heart beats about 2.5 billion times during an average life span. A heartbeat is made up of a contraction and a relaxation, just like other muscles in the body. This muscular pump, however, beats automatically. With each contraction of the heart, blood is pumped to the many parts of the body. Use the terms in the word box to label the diagram.

left atrium	right atrium	left ventricle	right ventricle
left lung	right lung	upper body	lower body

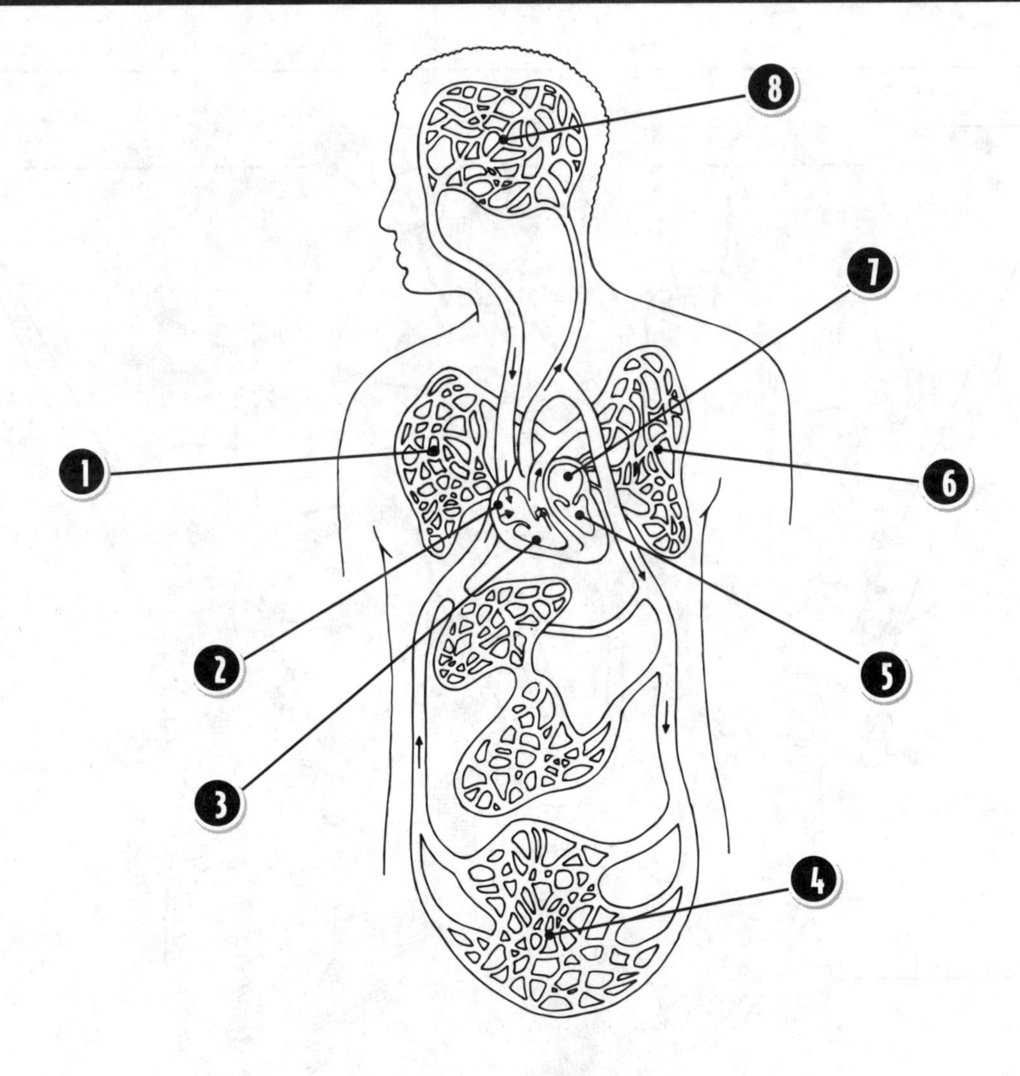

1 ______________________ 2 ______________________

3 ______________________ 4 ______________________

5 ______________________ 6 ______________________

7 ______________________ 8 ______________________

Name ______________________________ Date ______________

What's in Your Blood?

Blood is the fluid that is transported by the circulatory system. Match each term in the word box to its description.

plasma	red blood cells	white blood cells	platelets	leukocyte
water	fibrinogen	blood volume	4–5 quarts	hemoglobin

1. ______________________________ These patrol tissues where they get rid of damaged or dead cells and anything recognized as foreign to the body.
2. ______________________________ Another name for a white blood cell.
3. ______________________________ These release substances that start the process of blood clotting.
4. ______________________________ This refers to the amount of blood in a body, averaging 6 to 8 percent of total body weight for an adult human.
5. ______________________________ Mostly water, this functions as the medium through which blood cells and platelets are transported through the body.
6. ______________________________ These are concave disks with a squashed-in center. They transport oxygen and carry away some carbon dioxide.
7. ______________________________ On average, this is the volume of blood contained in an adult human.
8. ______________________________ This is an iron-containing pigment that gives red blood cells their color.
9. ______________________________ 90 percent of plasma is made up of this solvent.
10. ______________________________ A protein in blood plasma that is necessary for the clotting of blood.

Name ________________________________ Date ______________

Go with the Flow, Part I

The blood is pumped from one area of the circulatory system to the other. Use the terms in the word box to complete each description. Refer to the diagram on page 33 to help you trace the path of blood.

ventricle	right valve	pulmonary	right atrium	valve
vein	left atrium	left ventricle	aorta	inferior vena cava

1. Blood enters the ________________ from the superior vena cava and the interior vena cava.
2. Fluid pressure opens the atrioventricular ________________, allowing blood to enter the next chamber.
3. Blood flows into the right ________________. As pressure in this chamber rises, it forces the valve shut.
4. The right semilunar ________________ is forced open. Blood leaves the heart.
5. The blood flows through the ________________ artery to the lungs where it will transfer carbon dioxide and become rich with oxygen.
6. The blood, now rich in oxygen, re-enters the heart through the pulmonary ________________.
7. The oxygenated blood first enters the ________________.
8. Again, the pressure builds and the blood flows through the left atrioventricular valve into the ________________.
9. The ventricle contracts, pushing the blood through the left semilunar valve into the ________________.
10. Through this large artery the blood is transported to the capillary beds. The blood eventually returns to the heart through the superior and ________________. The cardiac cycle begins all over again.

Name ____ Date ________________

Go with the Flow, Part II

Use the steps on page 32 to label the path of blood through the heart. A starting point has been determined with the number 1. Number the remaining circles, and their corresponding parts, to show the path of blood through the heart.

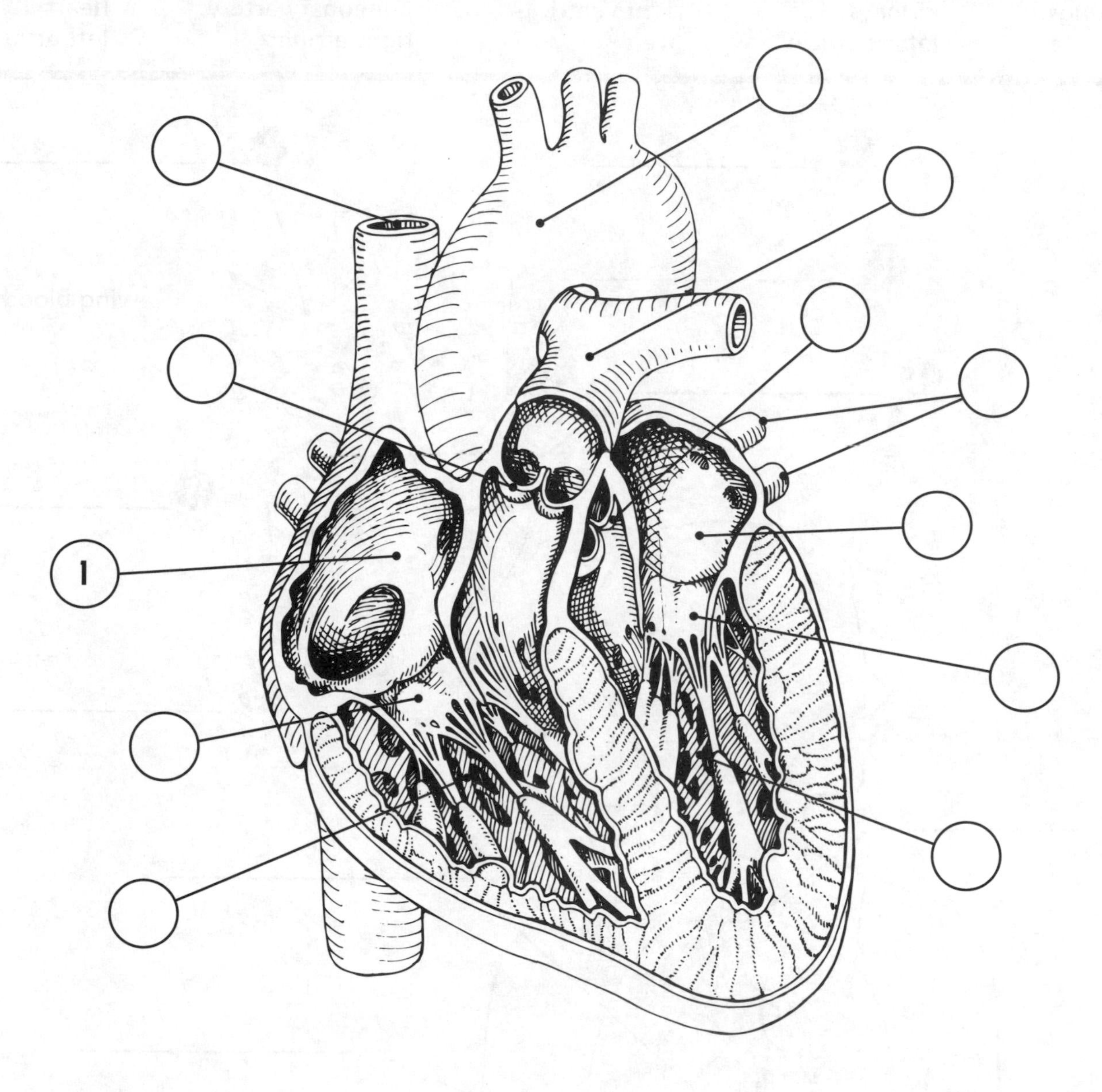

__1__ right atrium

_____ left atrioventricular valve

_____ pulmonary vein

_____ right ventricle

_____ right atrioventricular valve

_____ left ventricle

_____ aorta

_____ inferior vena cava

_____ pulmonary artery

_____ left semilunar valve

_____ left atrium

_____ right semilunar valve

Name ______________________ Date ______________

Circulatory Review

Use the terms in the word box to label each diagram.

lungs	kidneys	right ventricle	pulmonary artery	heart
aorta	left ventricle	liver	right atrium	left atrium

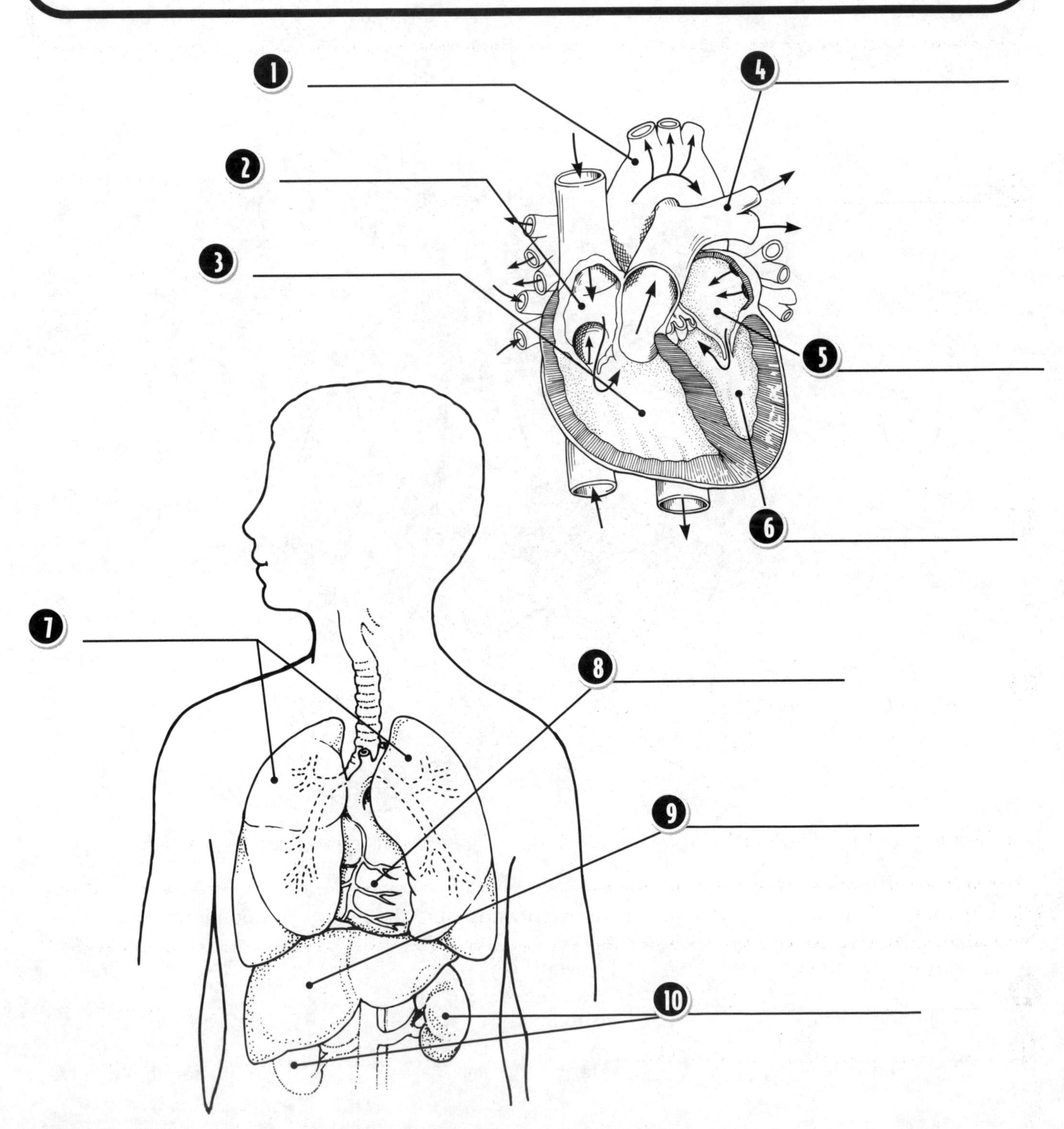

Name ____________________ Date ____________

The Respiratory System

The respiratory system is closely linked to the circulatory system. While the circulatory system transports oxygen to and carbon dioxide from the cells, it is the respiratory system that adds oxygen to the blood and removes carbon dioxide from the body. Use the terms in the word box to label the diagram.

pharynx	trachea	larynx	pleura
diaphragm	bronchial tube	nasal cavity	mouth

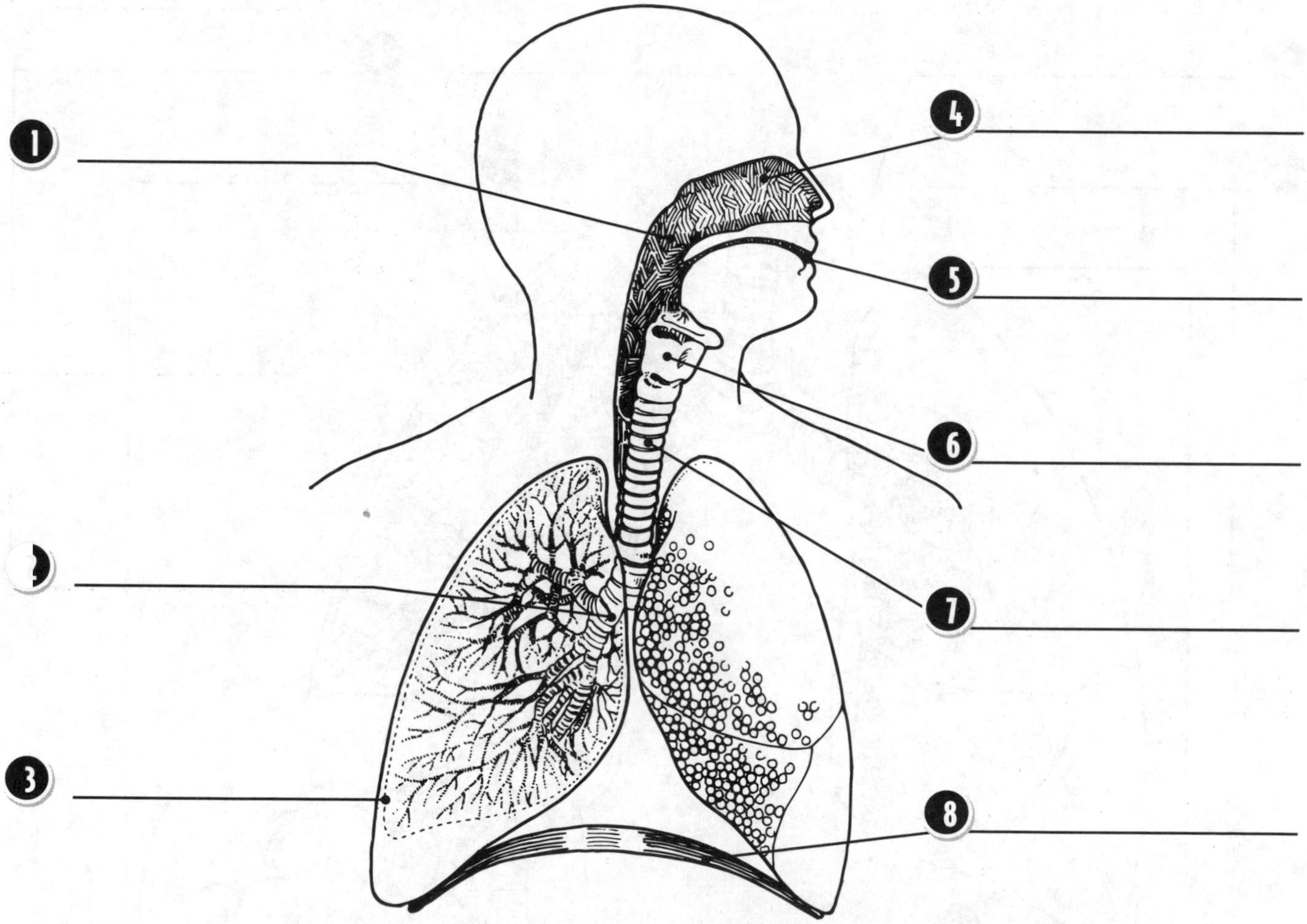

1. ____________________
2. ____________________
3. ____________________
4. ____________________
5. ____________________
6. ____________________
7. ____________________
8. ____________________

Use the terms in the word box below to match the common names to the scientific names.

throat	windpipe	voice box	lung cover

9. ____________________ larynx
10. ____________________ pleura
11. ____________________ trachea
12. ____________________ pharynx

Name ______________________________ Date ______________

The Lungs

The lungs are the main organ of the respiratory system. They are large spongy organs that fill a large part of the chest cavity. At birth the lungs are pink, but over time they become gray and mottled from tiny particles we breathe in with the air. Use the terms in the word box to label the diagram.

trachea	pleura	bronchiole	right lung	left lung
lobe	bronchial tube	alveoli	capillaries	alveolar sac

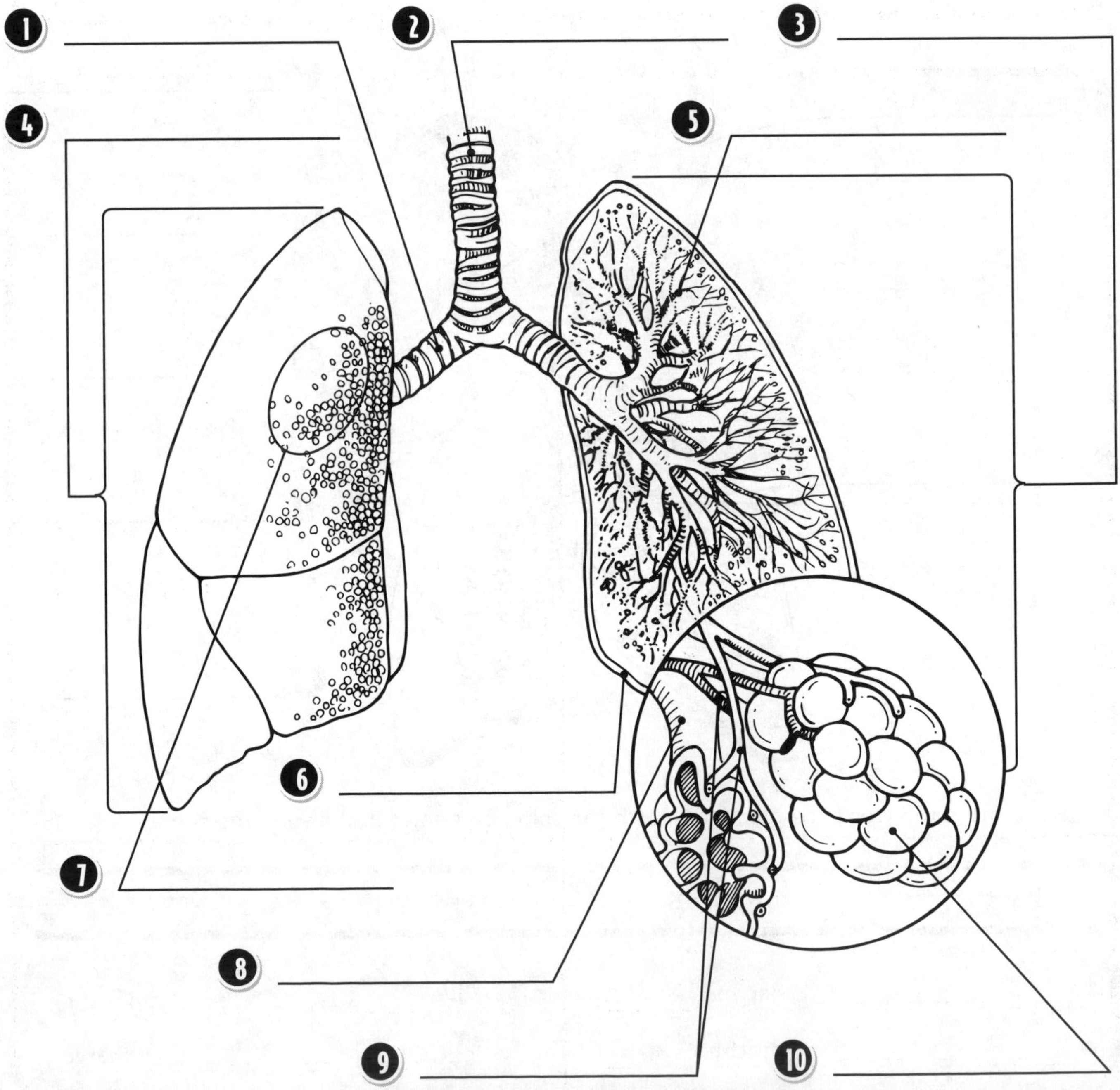

1 ______
2 ______
3 ______
4 ______
5 ______
6 ______
7 ______
8 ______
9 ______
10 ______

Name ______________________________ Date ______________

Breathe In, Breathe Out

Are breathing and respiration the same? Breathing is the process of bringing air rich in oxygen into and out of the lungs. Respiration is all of the processes that get oxygen to the cells, including breathing, movement of oxygen from lungs into the blood, transport by the blood, and movement of the oxygen from the blood into the cells. Use the phrases in the word box to complete the chart.

diaphragm is contracted	diaphragm is relaxed	oxygen goes in
carbon dioxide goes out	rib cage expands	diaphragm flattens
diaphragm moves up	rib cage returns to resting position	

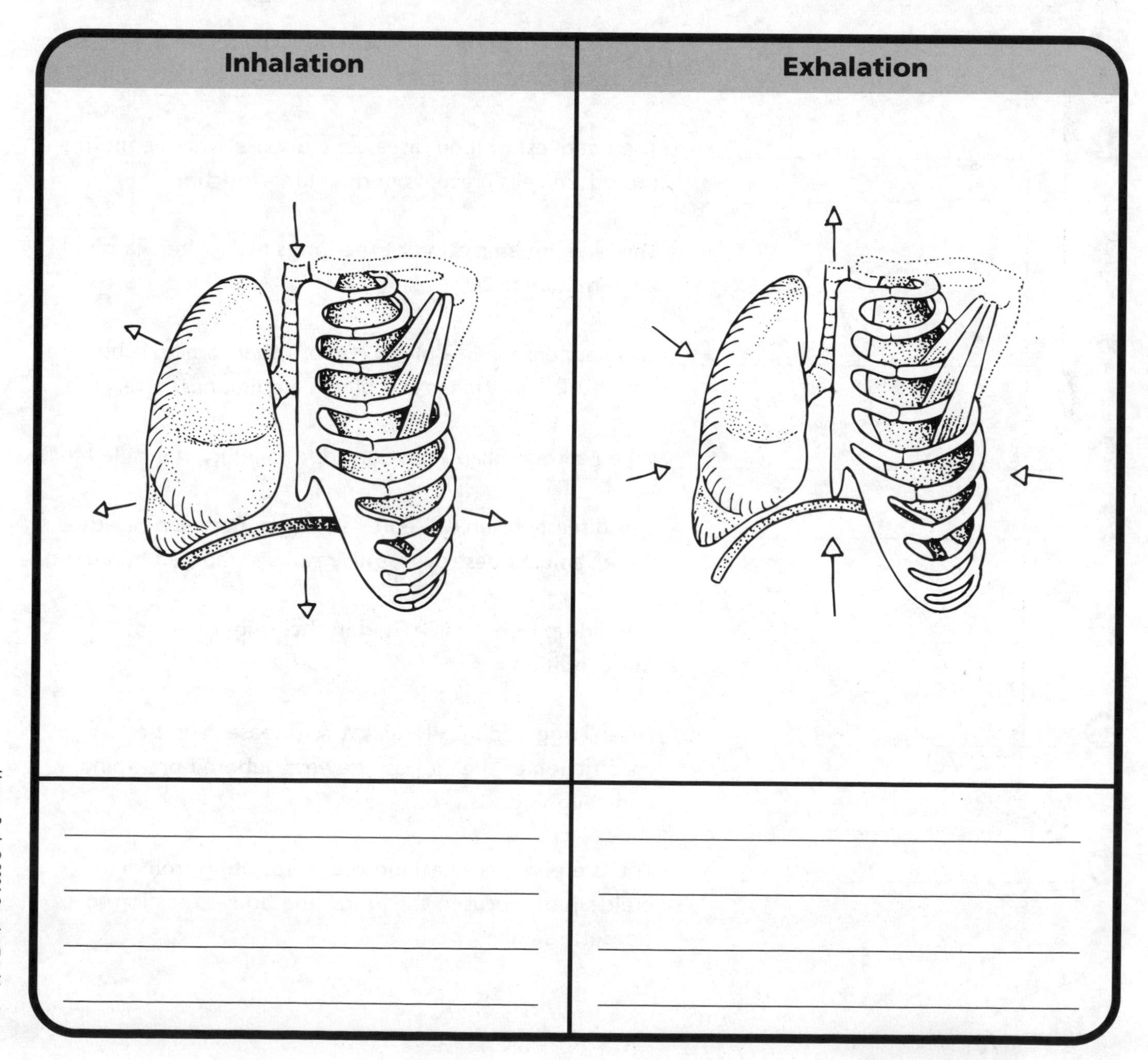

Name ______________________________ Date ______________

When Lungs Break Down

The respiratory system can be damaged from exposure to bacteria, viruses, and injury. It can also be damaged, sometimes with deadly consequences, by the choices humans make. The effects of smoking result in many of the diseases on this list. Match each lung condition in the word box to its description.

bronchitis	emphysema	asthma	lung cancer
smoker's cough	pneumonia	pneumothorax	pulmonary edema
pulmonary embolism			

1. ______________________ A disease caused by inflammation of the mucus membranes inside the bronchial tubes.

2. ______________________ The deadliest of lung diseases, a disease where abnormal growth of cells prevents normal lung function.

3. ______________________ This is an inflammation of the lungs most often caused by bacteria or viruses.

4. ______________________ This happens when a blood clot or other foreign substance gets stuck in the lungs and blocks a pulmonary artery.

5. ______________________ When a lung collapses, usually due to injury, it is called this.

6. ______________________ This disease, often caused by smoking, happens because the alveoli are destroyed. This results in labored breathing.

7. ______________________ A build-up of excessive fluid in the lungs leads to this condition.

8. ______________________ Often triggered by allergies, this disease causes a constriction of the air passageways, labored breathing, and coughing.

9. ______________________ This is a persistent hacking cough resulting from a build-up of mucus in the lungs, the body's reaction to cigarette smoke.

Name ____________________ Date __________

Respiratory Review

Classify the terms in the word box into the three categories on the chart.

bronchitis	alveoli	bronchial tube	trachea
emphysema	asthma	pneumothorax	diaphragm
pleura	bronchiole	lobe	lung

Lung Disorders		____________________ ____________________ ____________________ ____________________
Parts of a Lung		____________________ ____________________ ____________________ ____________________
Respiratory System		____________________ ____________________ ____________________ ____________________

Name ______________________________ Date ______________

The Digestive System

The purpose of the digestive system is to break down, or digest, the food we eat. This system is made up of a series of connected organs and tubes. Use the terms in the word box to label the diagram.

pancreas	liver	mouth	anus
teeth	stomach	esophagus	salivary glands
gallbladder	colon	small intestine	tongue

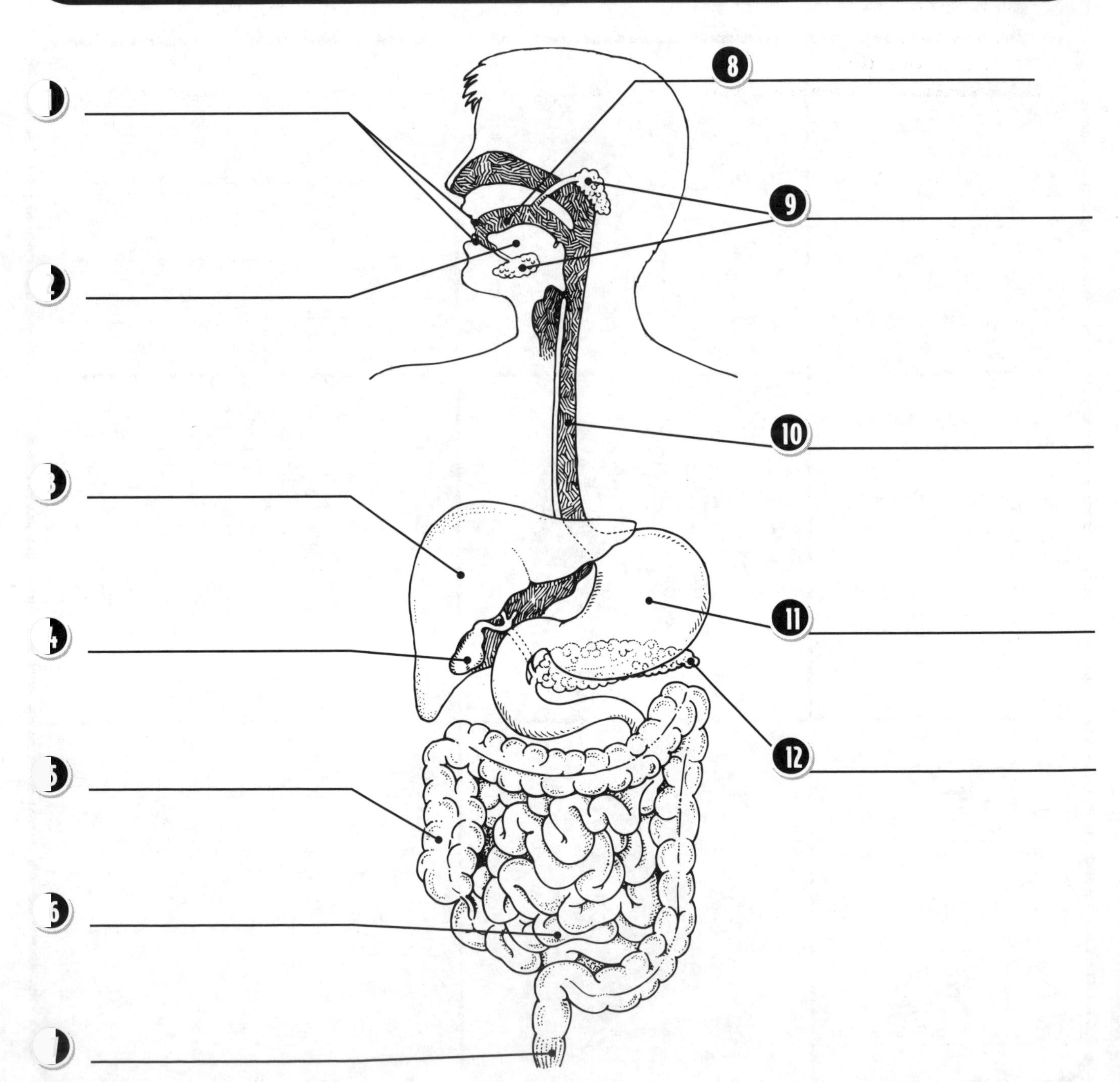

Name ______________________ Date ______________

The Alimentary Canal

The main part of the digestive system is the alimentary canal. This is a tube that begins with the mouth at one end and ends with the anus at the other end. Use the text in the word box to label the diagram.

anus	mouth	small intestine
large intestine	esophagus	stomach

food enters	water passes into the bloodstream
solid waste exits	pancreatic enzymes enter
bile enters	nutrients absorbed into bloodstream
squeezes food down to stomach	holds food while digesting

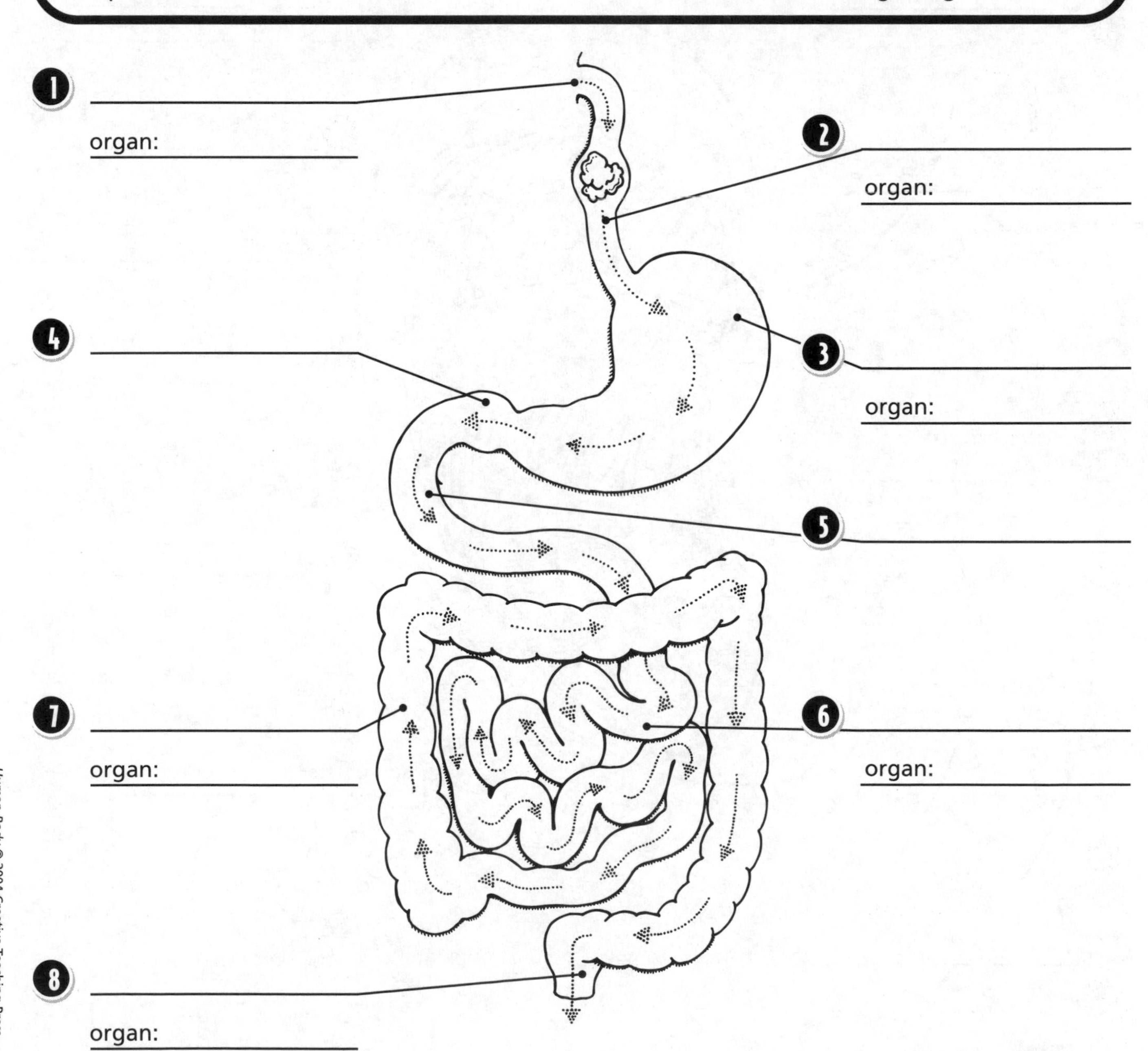

Name ______________________________ Date ____________

The Stomach

The stomach is a sack-like structure with muscular walls. It is the widest part of the alimentary canal. Three layers of muscle in the stomach wall allow it to contract in different directions. This motion mashes food and mixes it with digestive juices. Use the terms in the word box to label the diagram.

sphincter	serosa	longitudinal muscle	circular muscle
oblique muscle	mucous membrane	duodenum	esophagus

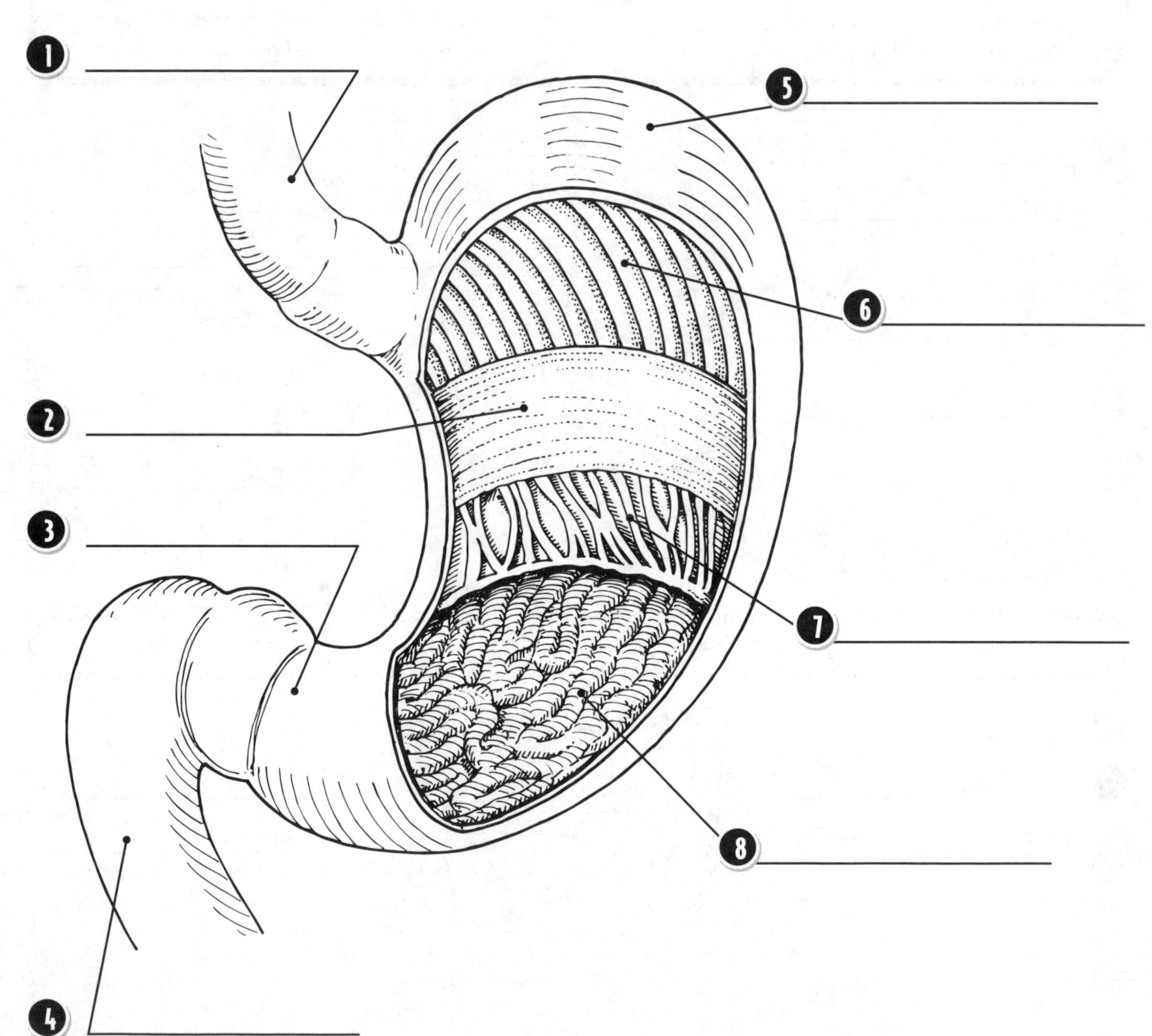

Name ______________________________ Date ______________

Digestion Helpers

In addition to the alimentary canal, there are other organs that help with the digestive process. Each produces fluids that help break down foods. Use the terms in the word box to label the diagram.

liver	bile duct	pancreas	duodenum	gallbladder

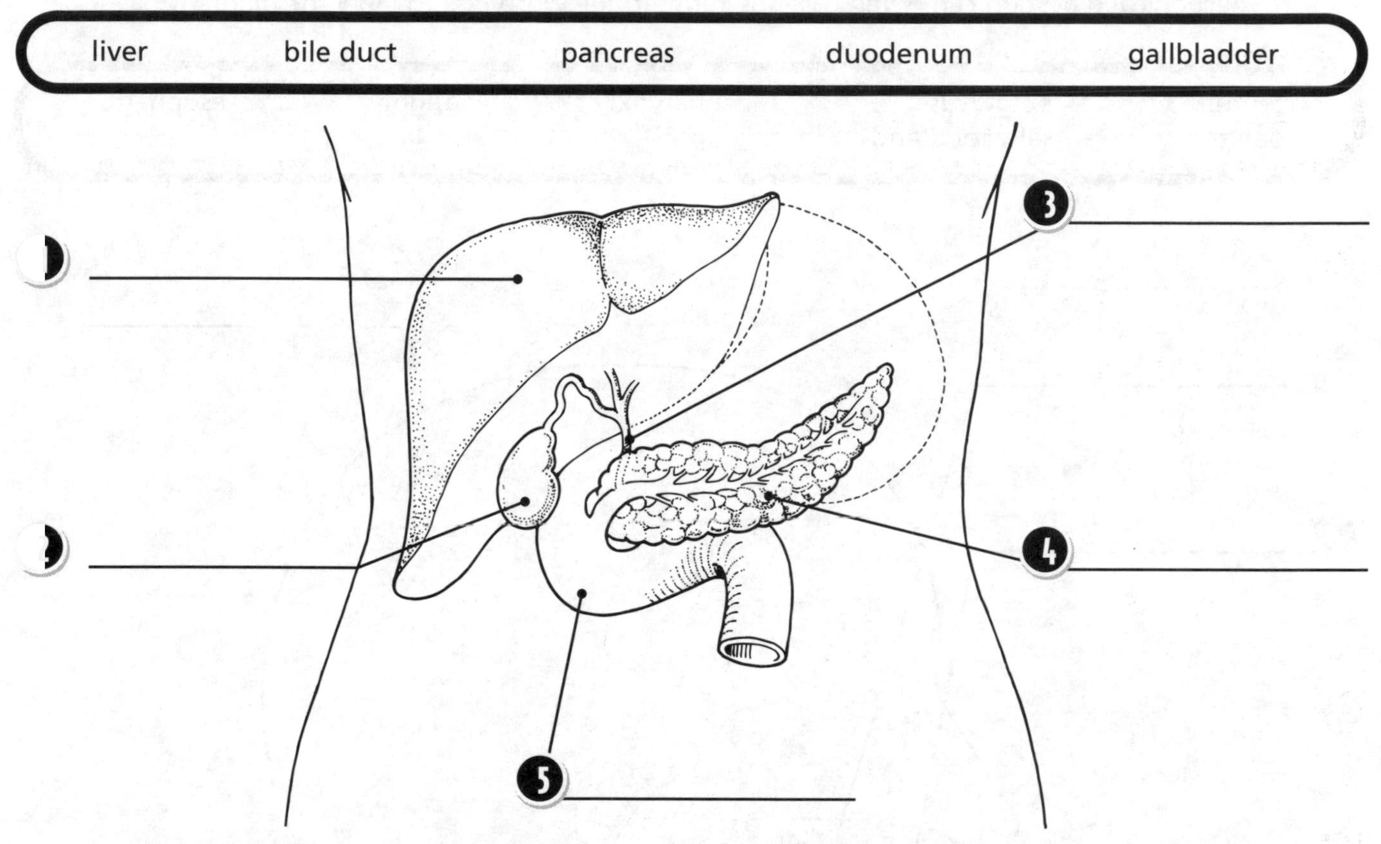

1. ______________________
2. ______________________
3. ______________________
4. ______________________
5. ______________________

Each "digestion helper" produces a fluid with a particular function. Match each term in the word box below with its description.

gallbladder	pancreatic juice	Brunner's glands	bile	enzymes

6. ______________________ The liver produces this fluid, which helps break down fats.

7. ______________________ When food is not traveling through the gut, bile is stored in this organ for later use.

8. ______________________ This fluid contains enzymes that break down sugars and starches into simple sugars, fats into fatty acids and glycerol, and proteins into amino acids.

9. ______________________ This organ secretes mucus to protect the stomach lining so it is not digested by all of these enzymes.

10. ______________________ These substances help start the reactions that break down food into parts that can be absorbed by the body.

Name ______________________________ Date ______________

Mouthing Off!

Did you know that digestion begins even before you put something in your mouth? The sight and smell of food triggers the salivary glands to produce fluid in anticipation of the food to be eaten. Once in the mouth, mechanical digestion can begin. Use the terms in the word box to label the diagram.

teeth	epiglottis	pharynx	tongue	esophagus
palate	salivary glands	lip	uvula	

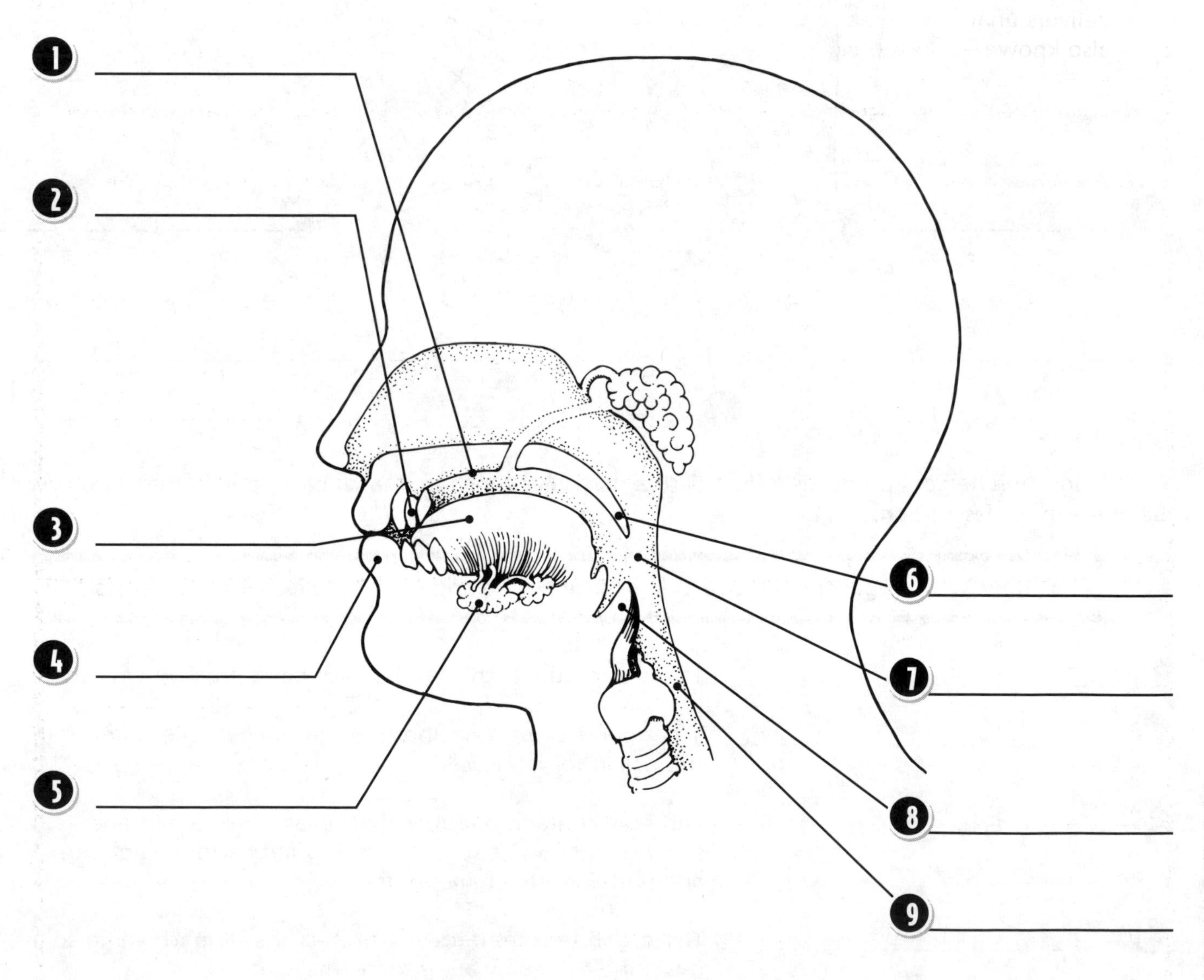

Name ____________________ Date __________

The Intestines

Use the information to complete the chart.

three parts: ascending, transverse, and descending
three parts: duodenum, jejunum, and ileum
stores undigested food
most nutrients are digested and absorbed
about 5 feet long
about 11 feet long
absorbs minerals and water
receives secretions from liver, pancreas, and gallbladder
delivers unabsorbed material to the colon
also known as the large intestine
where the appendix is attached

The Small Intestine

The Colon

Name ____________________ Date ____________

Digesting What You Have Learned

To review the digestive system, match each term in the word box to its description.

small intestine	anus	esophagus	pancreas	liver
stomach	colon	duodenum	tongue	salivary glands

1. ____________ This flexible organ of the mouth helps push food toward the teeth and aids in swallowing.

2. ____________ This organ produces bile for the digestion of fat.

3. ____________ This is a muscular tube that squeezes food down to the stomach.

4. ____________ This long organ removes water and stores solid waste until it is eliminated.

5. ____________ Final digestion takes place in this long tube.

6. ____________ This is the first portion of the small intestine. Food enters it after it leaves the stomach.

7. ____________ These produce a fluid that enters the mouth and moistens the food, beginning the digestion of some starches.

8. ____________ This is the opening at the end of the alimentary canal through which solid waste leaves the body.

9. ____________ This organ stores the food for 3 to 4 hours while digestion is occurring.

10. ____________ This organ secretes enzymes that help digest fats, sugars and starches, and proteins.

Name ______________________ Date __________

The Urinary System

The human body needs water to survive. The urinary system makes sure the body does not have too much or too little fluid at any given time. Use the terms in the word box to label the diagram.

vein	kidney	muscle	bladder	ureter	urethra	artery

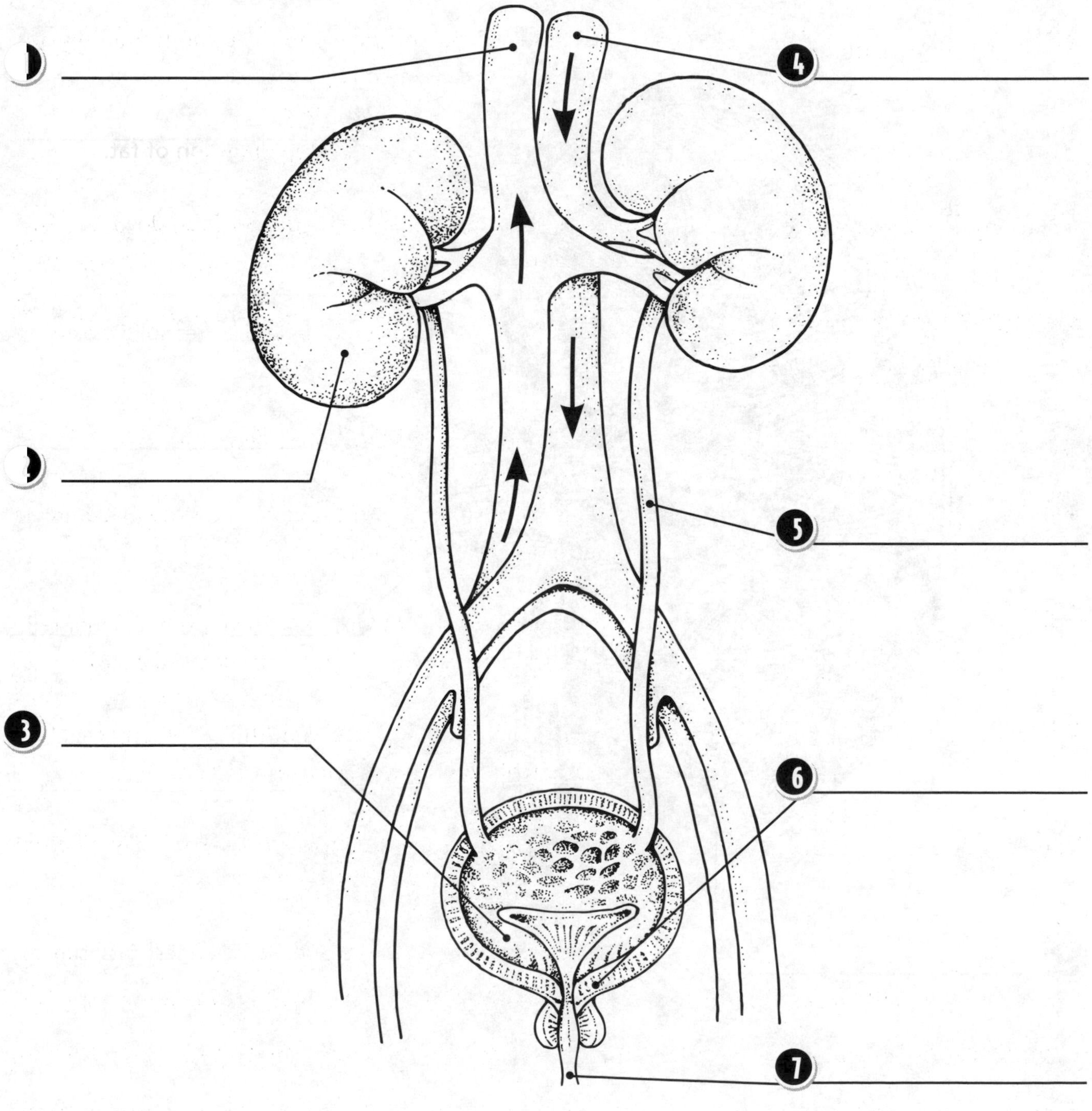

1 ______________________

2 ______________________

3 ______________________

4 ______________________

5 ______________________

6 ______________________

7 ______________________

Name ______________________________ Date ______________

Two of a Kind

The kidneys are organs that filter water, minerals, organic wastes, and other waste substances out of the blood. The human body contains two kidneys. These bean-shaped organs are about as big as a fist. Use the terms in the word box to label the diagram.

renal artery renal vein ureter renal capsule kidney medulla kidney cortex

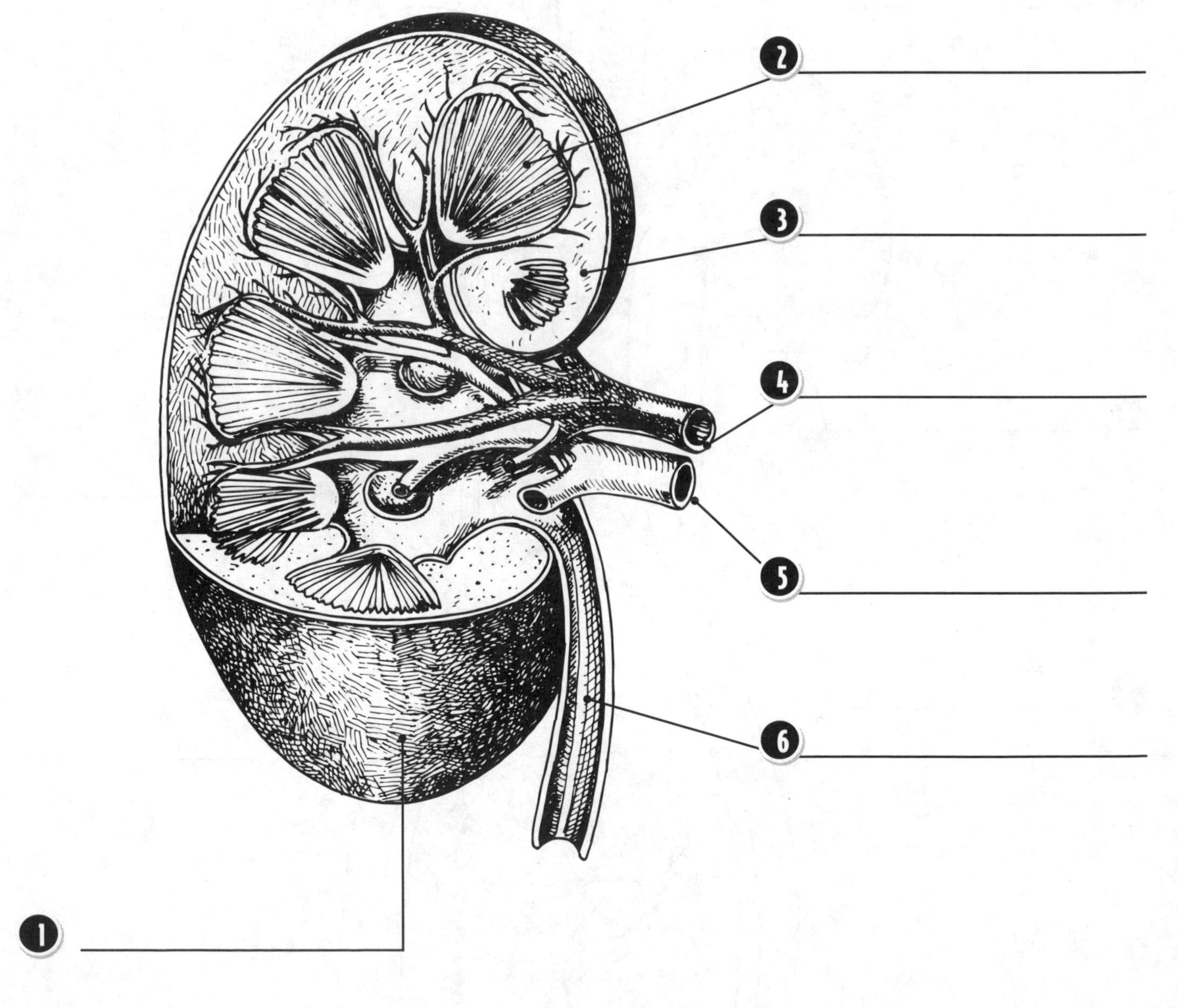

Name ______________________________ Date ______________

Maintaining a Balance of Fluids

Humans continually gain and lose water and dissolved substances called solutes. The urinary system maintains the proper level of fluid and solutes in the body at all times. Match each term in the word box to its description.

interstitial fluid	urine	extracellular fluid	hypertension
kidney stones	dialysis	thirst behavior	blood

1. ______________________ When the amount of fluid in the blood drops and the level of solutes rises, this response is triggered to seek out water.

2. ______________________ A fluid that transports substances to and from all tissues by way of the circulatory system.

3. ______________________ A fluid formed when the kidneys filter water and solutes from the blood. The excess water and solutes remains as this fluid.

4. ______________________ The fluid that is found in the spaces between cells and tissues.

5. ______________________ Hard deposits of uric acid, calcium salts, and other wastes that can become lodged in the ureter or urethra and block urine flow.

6. ______________________ All of the fluid not inside cells, it includes plasma and interstitial fluid.

7. ______________________ The filtering of fluids with a machine to remove and add solutes in proper concentration. This process is used when kidneys fail to function on their own.

8. ______________________ Abnormally high blood pressure that can happen when the amount of sodium in the body becomes too great.

Name ______________________________ Date ______________

The Lymphatic System

The lymphatic system helps transport digested fat from the intestine to the bloodstream. It also removes and destroys toxic substances and helps the body resist the spread of disease. Use the terms in the word box to label the diagram.

lymphatic duct	tonsils	thymus gland	bone marrow
spleen	lymph nodes	lymph vessels	thoracic duct

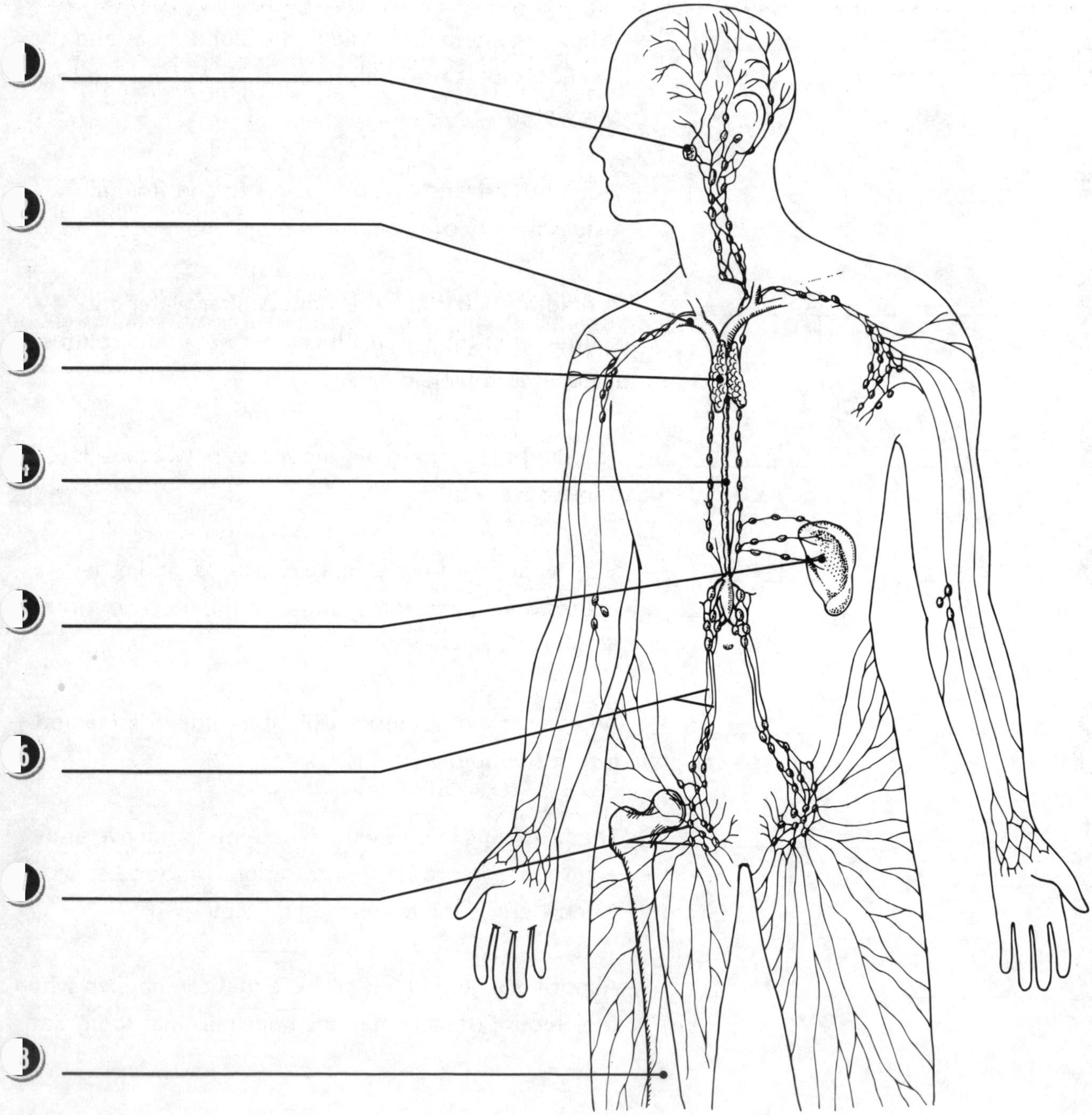

1 ______________________

2 ______________________

3 ______________________

4 ______________________

5 ______________________

6 ______________________

7 ______________________

8 ______________________

Name ______________________ Date ____________

Functions of the Lymphatic System

The lymphatic system is partly a system of tubes that collect and transport water and other solutions from the fluids between tissues and parts of organs to the circulatory system. In addition, other parts of the system filter out bacteria and help fend off illness. Match each term in the word box to its description.

tonsils	lymphatic ducts	thymus gland	thoracic duct
spleen	lymph vessels	lymph nodes	bone marrow

1. ______________ A small mass of tissue located in the throat between the mouth and the pharynx; believed to help protect against infections of the respiratory system.
2. ______________ A small organ positioned behind the top of the breastbone, it increases in size gradually from birth until puberty, after which it begins to shrink.
3. ______________ A large organ positioned to the left of the stomach and below the diaphragm, it stores blood, disintegrates old blood cells, and filters out foreign substances in the blood.
4. ______________ Produces white blood cells, which are then transported by the lymphatic system.
5. ______________ Any of the small bodies located along the lymphatic vessels, particularly in the neck, armpit, and groin. They filter bacteria and foreign particles from lymph fluid.
6. ______________ A system of tubes by which lymph fluid is transported throughout the body.
7. ______________ The main duct or vessel of the lymphatic system.
8. ______________ These vessels drain lymph from the upper portion of the body.

Name ______________________ Date __________

Working as a Team

The lymphatic system works together with the circulatory and urinary systems to maintain fluid levels and help the body stay healthy. How well do you know the parts and processes of each system? Use the text in the word box to complete the chart.

spleen heart kidney nodes capillaries ureters
white blood cells lymphocytes urine blood lymph

transports oxygen and nutrients
transports fats and pathogens
triggers thirst behavior
filters water and solutes
produces red blood cells
releases lymphocytes
produces antibodies
maintains fluid balance
forms platelets for clotting

Lymphatic System	Circulatory System	Urinary System
______	______	______
______	______	______
______	______	______
______	______	______
______	______	______
______	______	______
______	______	______
______	______	______
______	______	______
______	______	______
______	______	______
______	______	______
______	______	
______	______	

Name ______________________________ Date ______________

The Endocrine System

The endocrine system includes organs or glands in the human body that are responsible for the control of various functions. Use the terms in the word box to label the diagram.

thyroid gland	pineal gland	adrenal gland	testes (male)
ovaries (female)	pituitary gland	pancreas	thymus

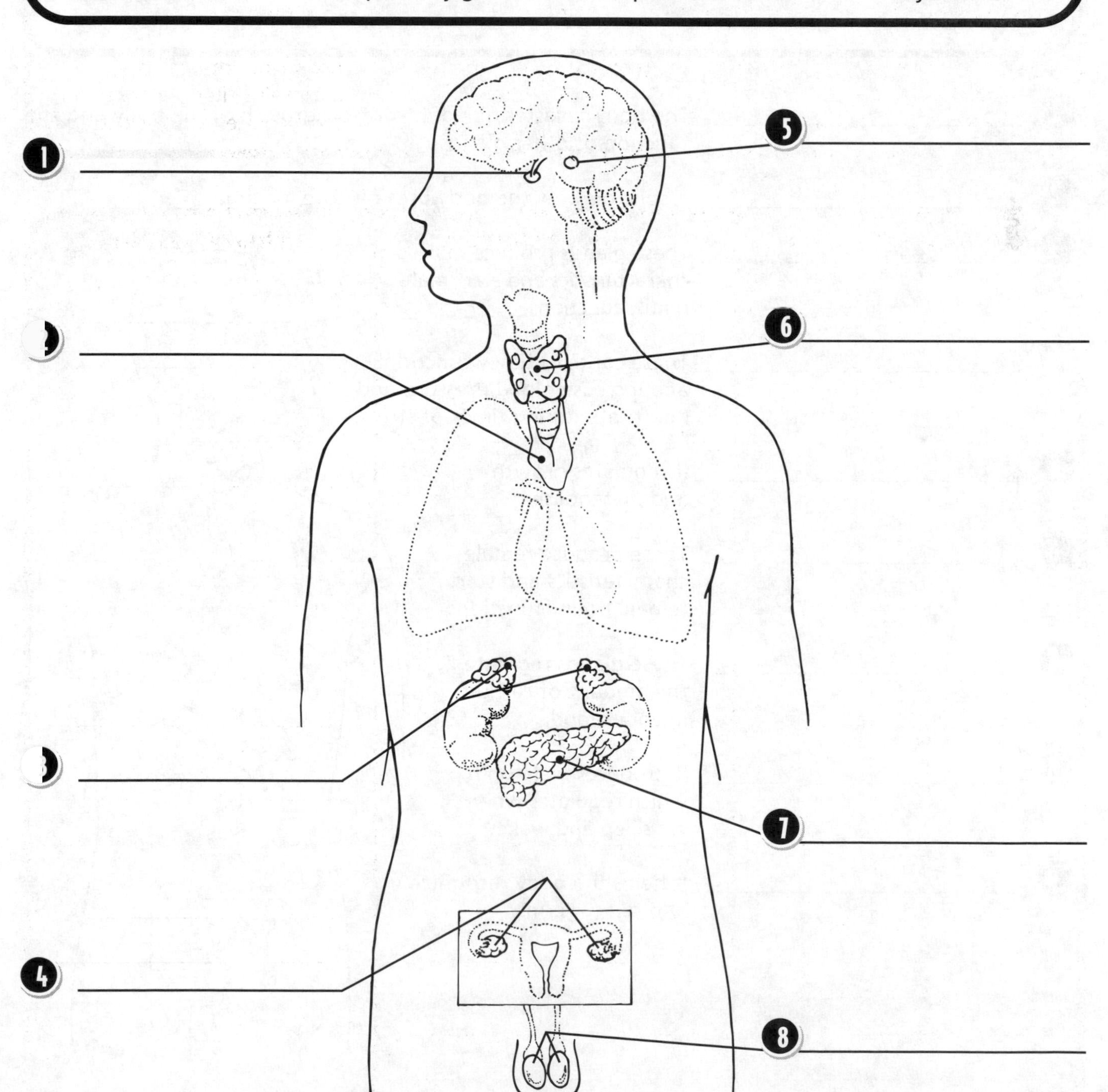

Name ______________________________ Date ______________

A Particular Gland for a Particular Function

Each gland in the endocrine system produces a hormone. Hormones are the body's chemical messengers. When hormones are released, they tell other parts of the body to do a specific job. Match each term in the word box to the function it names.

thyroid	pituitary	parathyroid	adrenal	thymus
ovaries	pancreas	testes	pineal gland	

1. ______________________ The body's master gland, it controls growth of the body and the function of other glands.
2. ______________________ It controls how the body uses glucose, or sugar.
3. ______________________ These glands produce male characteristics and start male bodily functions.
4. ______________________ These control kidney function and increase blood pressure and heart rate during times of stress.
5. ______________________ It regulates growth and metabolism.
6. ______________________ These produce female characteristics and start female bodily functions.
7. ______________________ These glands regulate the amount of calcium in your blood.
8. ______________________ It produces melatonin, which regulates when we sleep and wake.
9. ______________________ It helps the body recognize and reject germs.

Name ______________________________ Date ______________

Producing Hormones

Each gland in the endocrine system produces a specific hormone with a particular purpose. Use the text in the word box to complete the chart.

ovary	pancreas	adrenal gland	pineal gland
pituitary	parathyroid	testis	thyroid

regulates female functions	regulates male functions	regulates sleep-wake cycle
regulates glucose levels	regulates pain response	regulates calcium in blood
response to stress	regulates growth	

Name of Hormone	Gland	Function
1 estrogen		
2 melatonin		
3 insulin		
4 adrenaline		
5 thyroxine		
6 testosterone		
7 parathyroid hormone		
8 endorphins		

Name ______________________________ Date ______________

Where Do Those Hormones Come From?

Each hormone used to regulate functions in the human body is made by a certain gland in the body. Use each term in the word box to label the organ in the body that produces each hormone.

estrogen	melatonin	insulin	adrenaline
thyroxine	testosterone	parathyroid hormone	endorphins

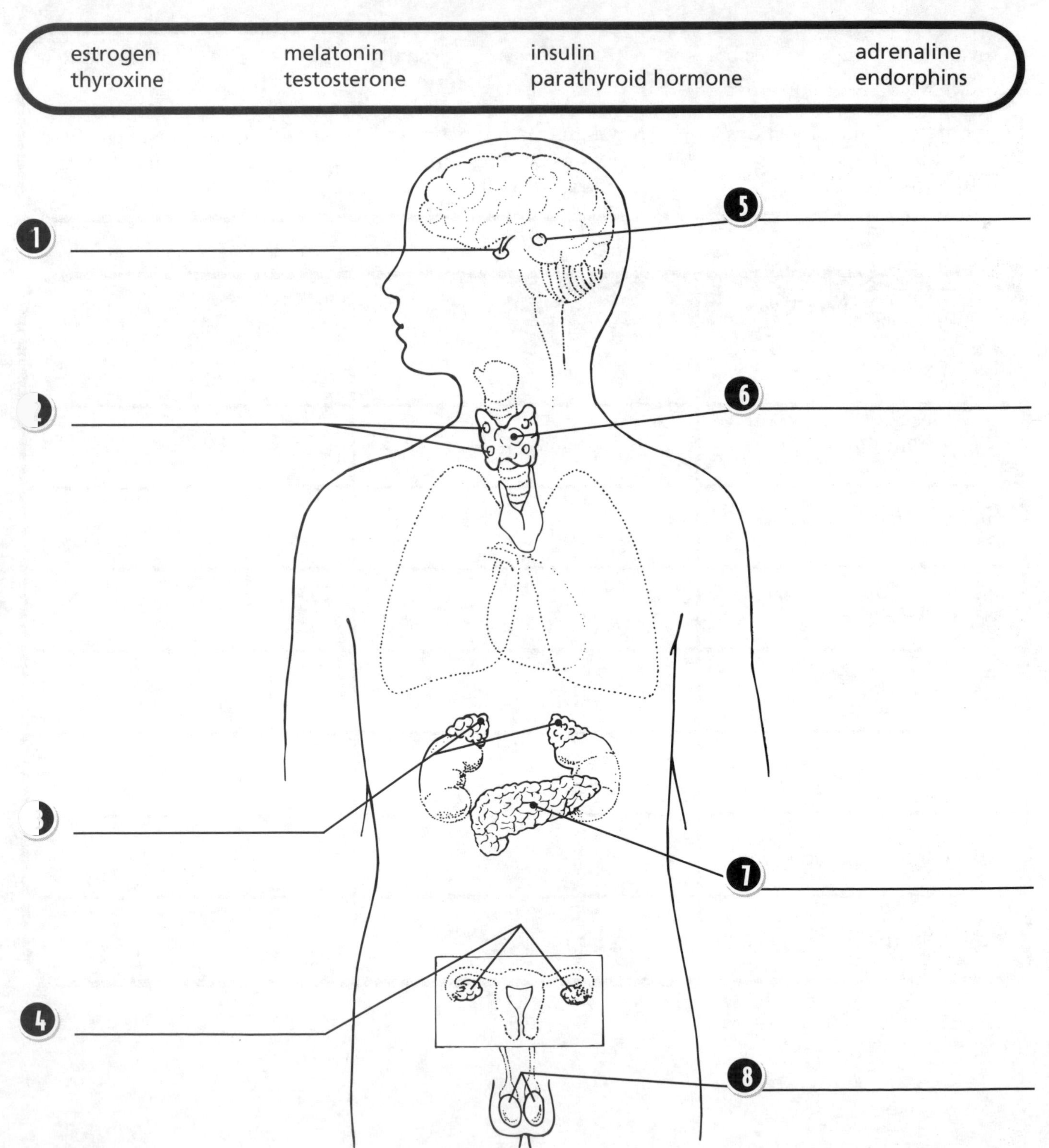

Name ______________________ Date ______________

The Central Nervous System

The central nervous system includes the brain and spinal cord. It processes and coordinates all information picked up by the senses and motor commands telling various body parts what to do. It is also the seat of brain functions such as memory, intelligence, learning, and emotion. Use the terms in the word box to label the diagram.

brain	cerebrum	nerve cell	spinal cord
nerves	cerebellum	brain stem	

1. ______________
2. ______________
3. ______________
4. ______________
5. ______________
6. ______________
7. ______________

Name ______________________________ Date ______________

Neurons

Neurons are nerve cells. They collect sensory information or changes in the environment. Then they activate muscles or glands to react. Use the terms in the word box to label the diagram.

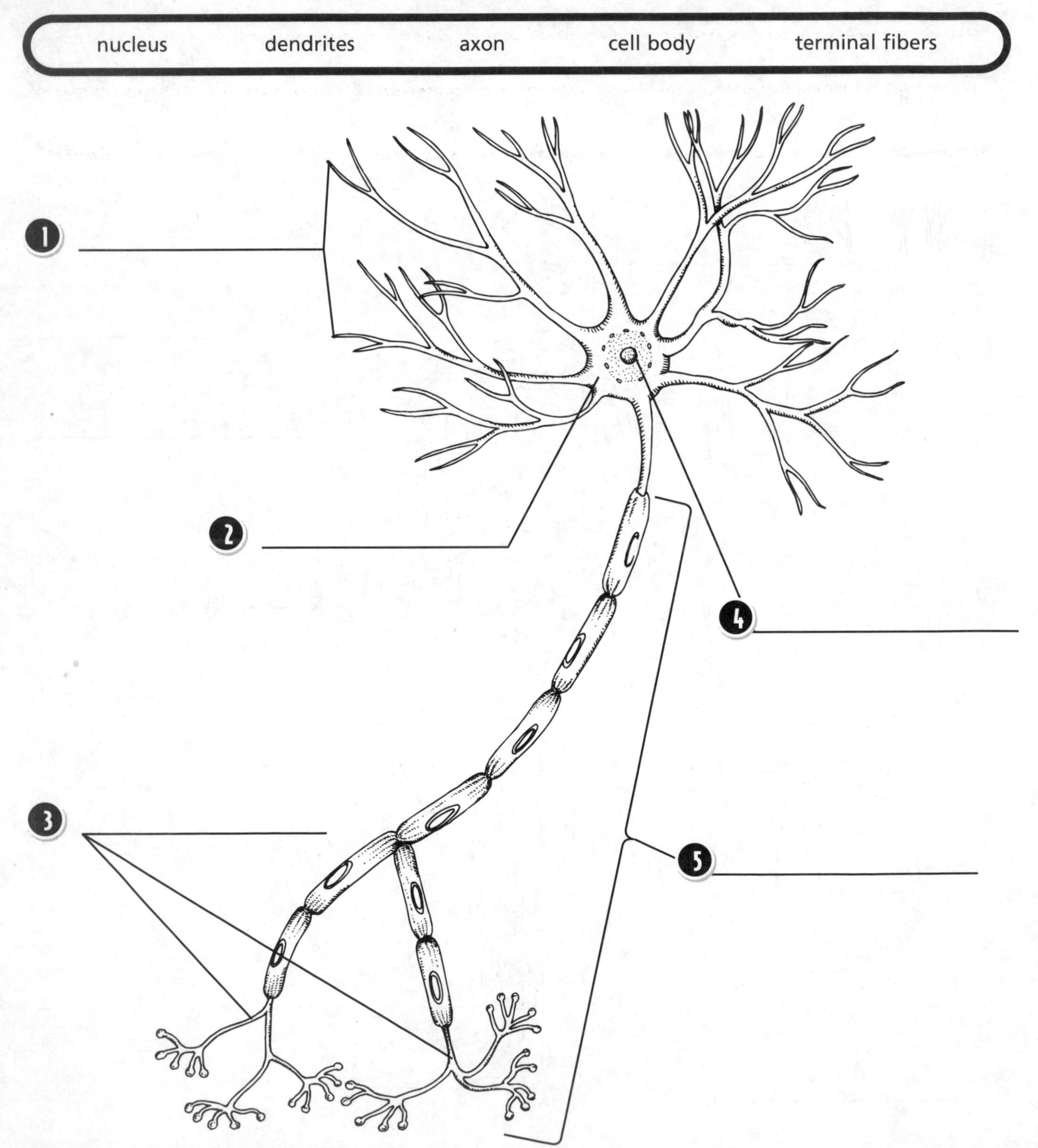

Name ______________________________ Date ______________

Impulse Transmitters

Neurons transfer information to and from one another through electrical impulses. These impulses pass from one neuron to another through connections called synapses. Use the terms in the word box to label the diagram.

synaptic cleft	axon	axon terminal	transmitting molecule	dendrite	synapse

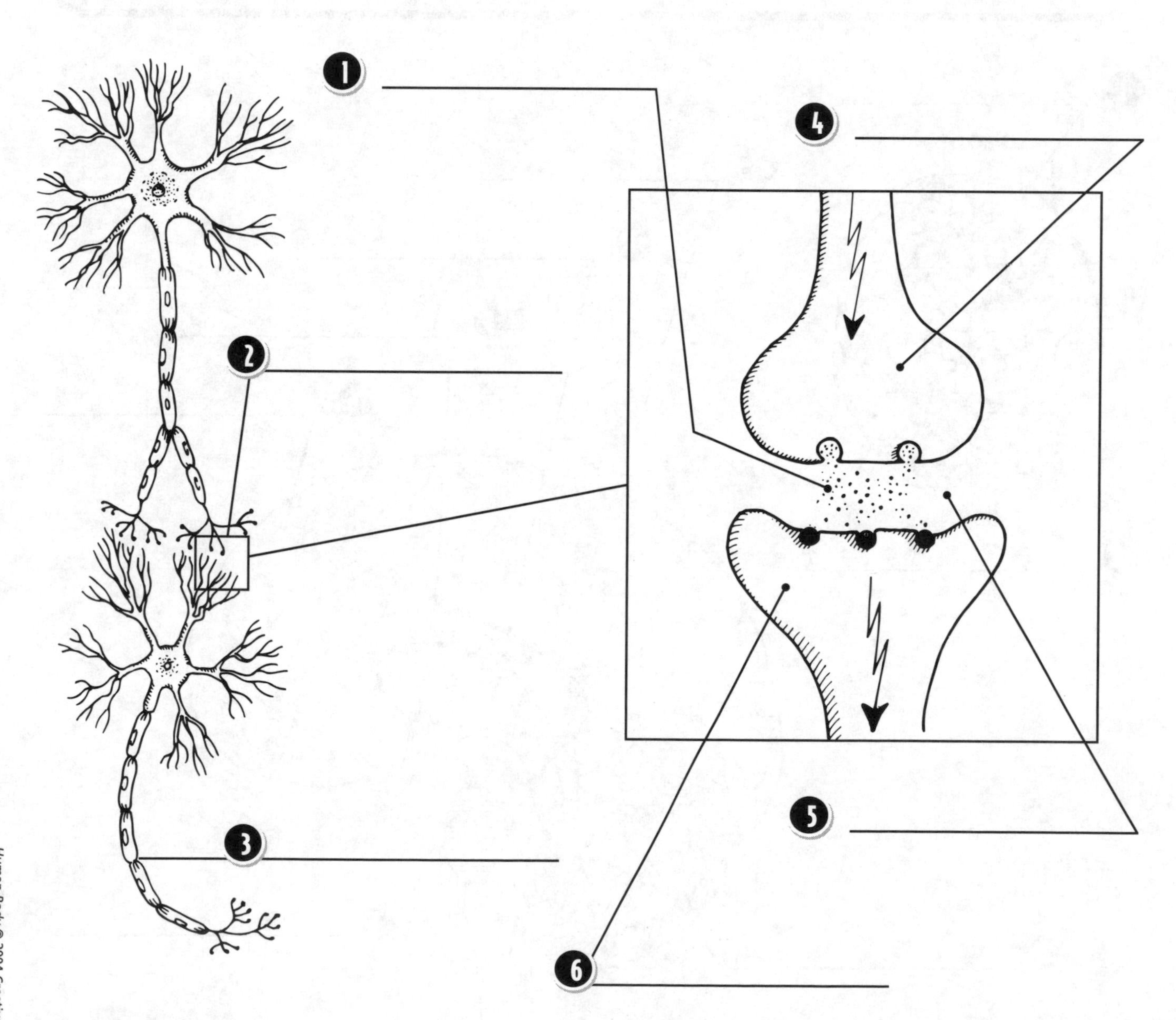

Name ______________________ Date __________

The Brain

The brain is a complex control center for the body. It can be divided into sections in a number of ways: into two hemispheres; into forebrain, midbrain, and hindbrain; and into lobes. However the brain is viewed, it is made up of three main parts: the cerebrum, the cerebellum, and the medulla. Use the terms in the word box to label the diagram.

cerebrum	cerebellum	medulla	left cerebral hemisphere	thalamus
hypothalamus	pineal gland	spinal cord	right cerebral hemisphere	

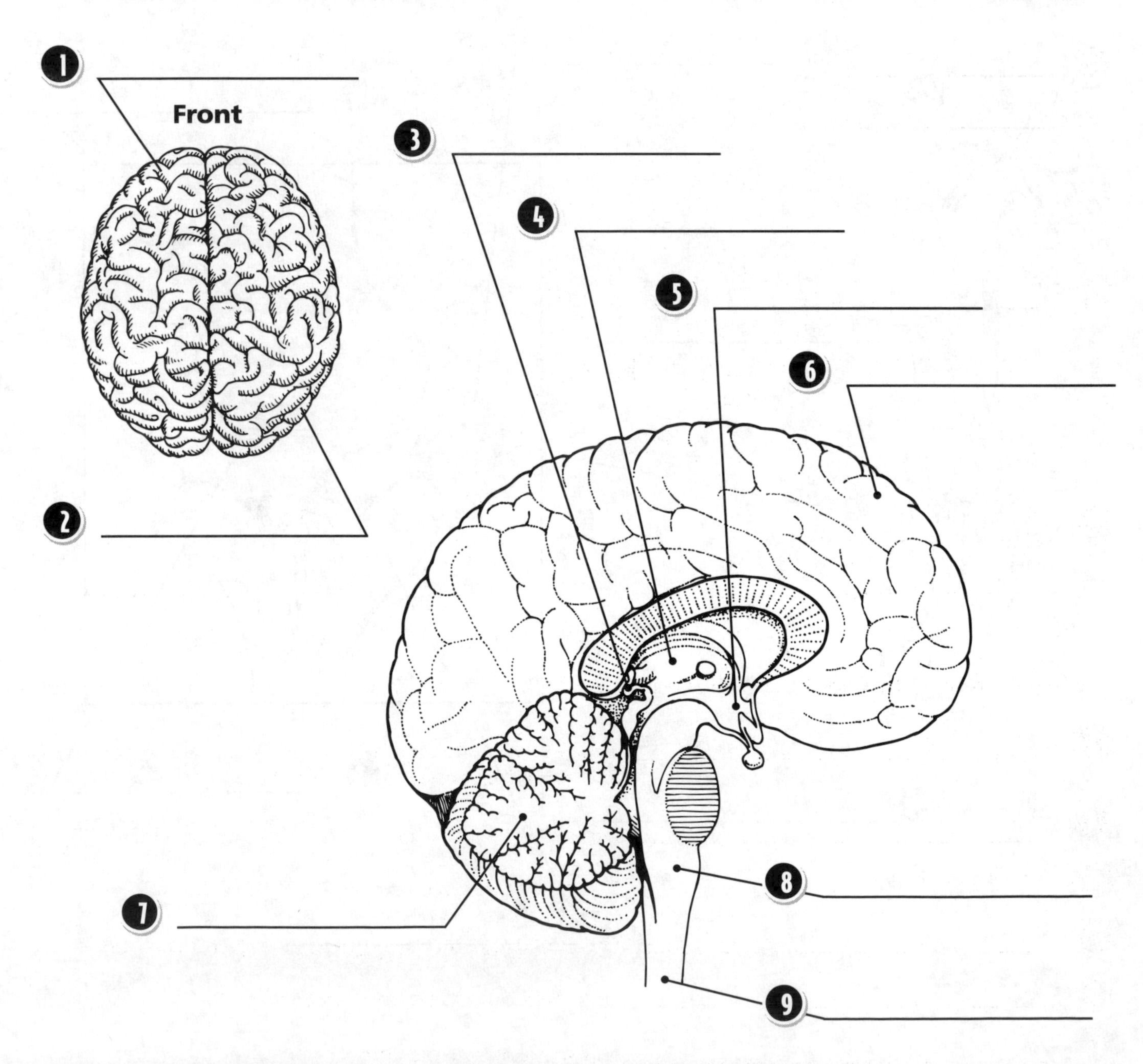

Name ______________________ Date ____________

The Lobes of the Brain

The cerebrum is the largest portion of the brain. This portion is responsible for most complex brain function. The cerebrum can be divided into lobes, each one having specific functions. Use the terms in the word box to label the diagram.

frontal lobe	motor cortex	somatosensory cortex
parietal lobe	temporal lobe	occipital lobe

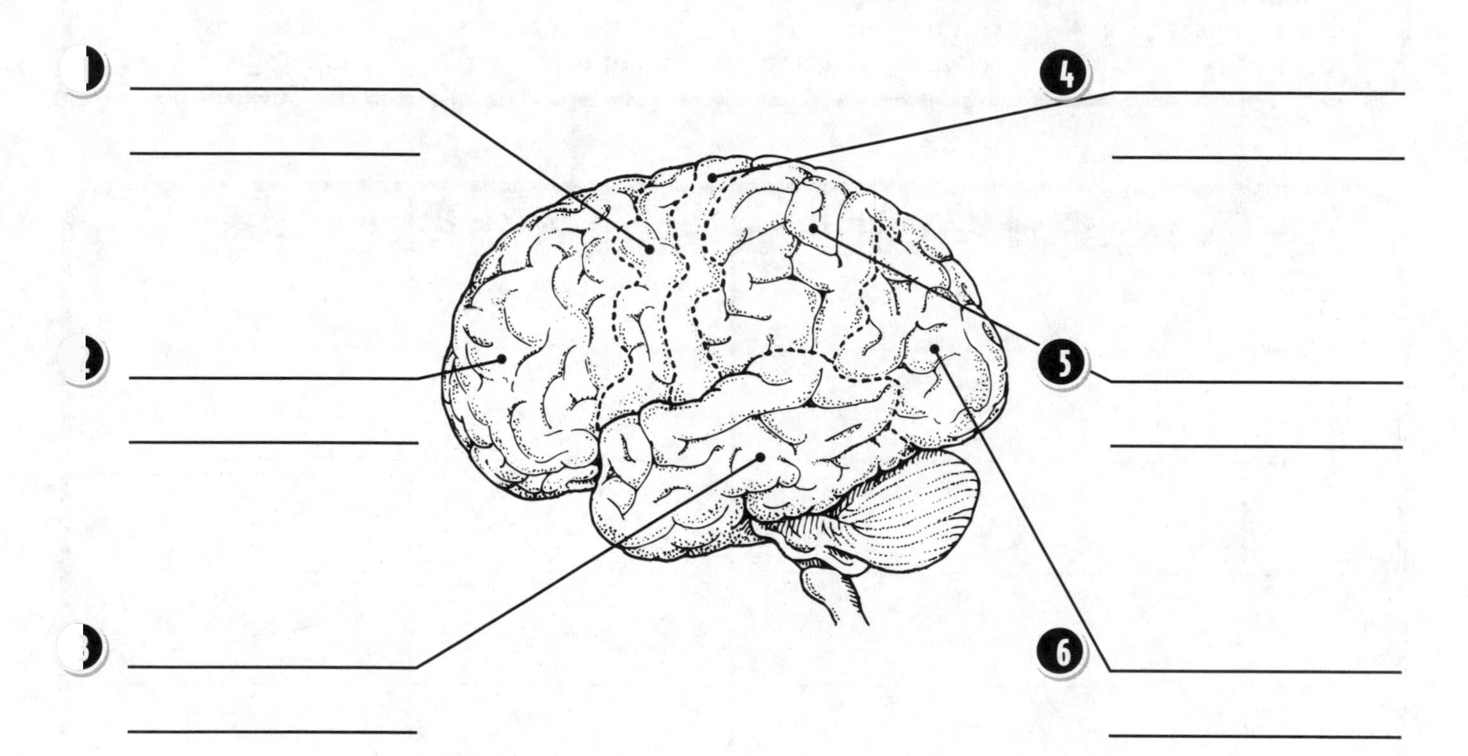

Match each term in the word box to its description.

7. ______________________ This section controls planning of movements, memory, and the inhibition of unsuitable behavior.

8. ______________________ This section controls hearing and visual processing.

9. ______________________ This section receives signals from the skin and joints.

10. ______________________ This section generates signals for motor responses.

11. ______________________ This section controls body sensations.

12. ______________________ This section controls vision.

Name ______________________________ Date ____________

Autonomic versus Somatic Nervous Systems

The autonomic and somatic nervous systems are controlled by the central nervous system. In the autonomic nervous system, nerves lead from the central nervous system to the smooth muscle, cardiac muscle, glands, organs, and other internal structures. The somatic nervous system connects the central nervous system to the skeletal muscles. It controls voluntary movements. Categorize the actions in the word box to complete the chart.

heart beats	fist clenches	fingers curl	pancreas secretes
produces bile	climb stairs	stomach digests	colon contracts
make a peace sign	curl tongue	get goosebumps	stretch arms
feel a hot pan	release adrenaline	point toes	diaphragm relaxes

Autonomic Nervous System	Somatic Nervous System
______________	______________
______________	______________
______________	______________
______________	______________
______________	______________
______________	______________
______________	______________
______________	______________

Name ______________________________ Date ____________

Autonomic Nervous System

The autonomic nervous system controls the life-sustaining functions of the body. Because of it, there is no need to think about keeping your heart beating or telling your stomach to digest food. The organs and muscle tissues of this system work involuntarily. Many important organs function due to their part in the autonomic nervous system. Use the terms in the word box to label the diagram.

eye	lungs	stomach	rectum	trachea	liver
pancreas	colon	heart	gallbladder	small intestine	

1. ______________________
2. ______________________
3. ______________________
4. ______________________
5. ______________________
6. ______________________
7. ______________________
8. ______________________
9. ______________________
10. ______________________
11. ______________________

Name ______________________ Date __________

Nervous System Functions

Each part of the central nervous system controls certain functions of the body. Injury to any part within this system can result in impairment of the corresponding function. Match each term in the word box to its function. Then write the number of the term and function in the corresponding circle to label the diagram.

cerebrum	cerebellum	medulla	spinal cord	spinal nerves

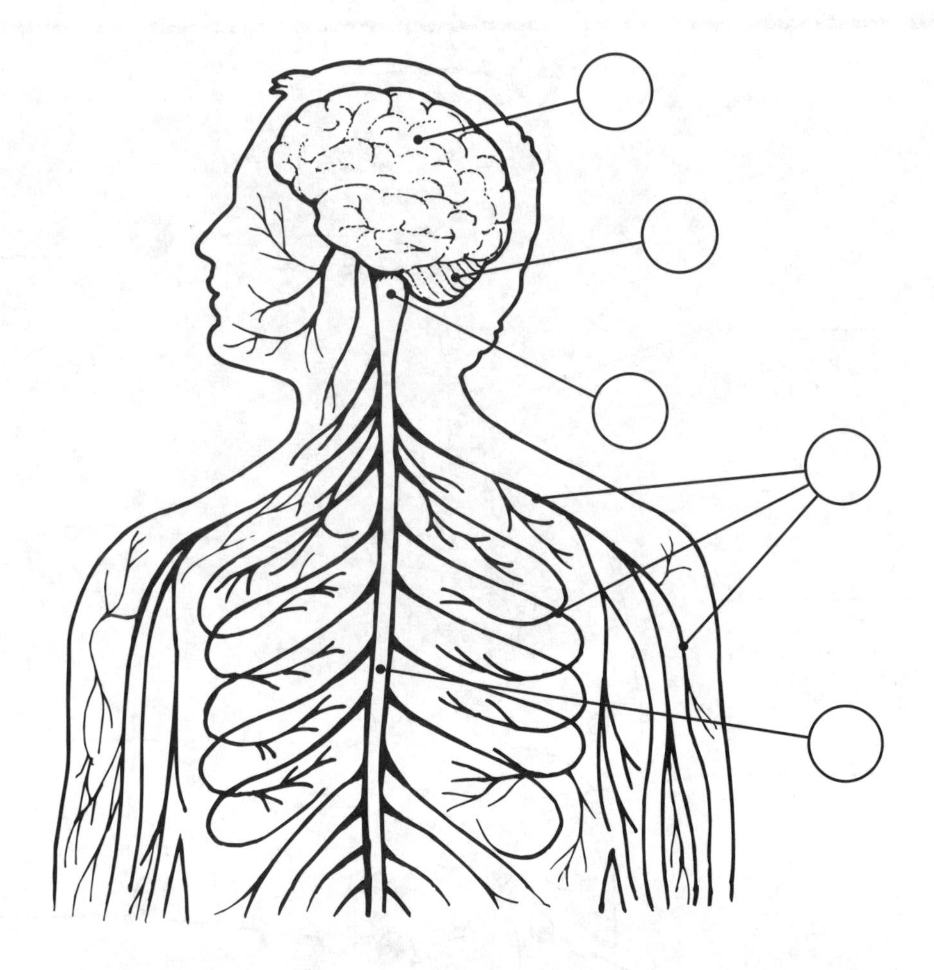

1. ______________ It controls balance and coordination of the muscles.
2. ______________ It controls breathing, heartbeat, and other vital processes within the body.
3. ______________ It controls thought, movement you choose to make, memory, and learning. It also processes information from the senses.
4. ______________ They carry impulses between the spinal cord and body parts.
5. ______________ It relays impulses between the brain and other parts of the body.

Name ______________________________ Date ______________

Central Nervous System Review

Use the terms in the word box to label the diagram.

cerebrum	cerebellum	medulla	spinal cord	axon
brain stem	dendrite	neuron	nucleus	

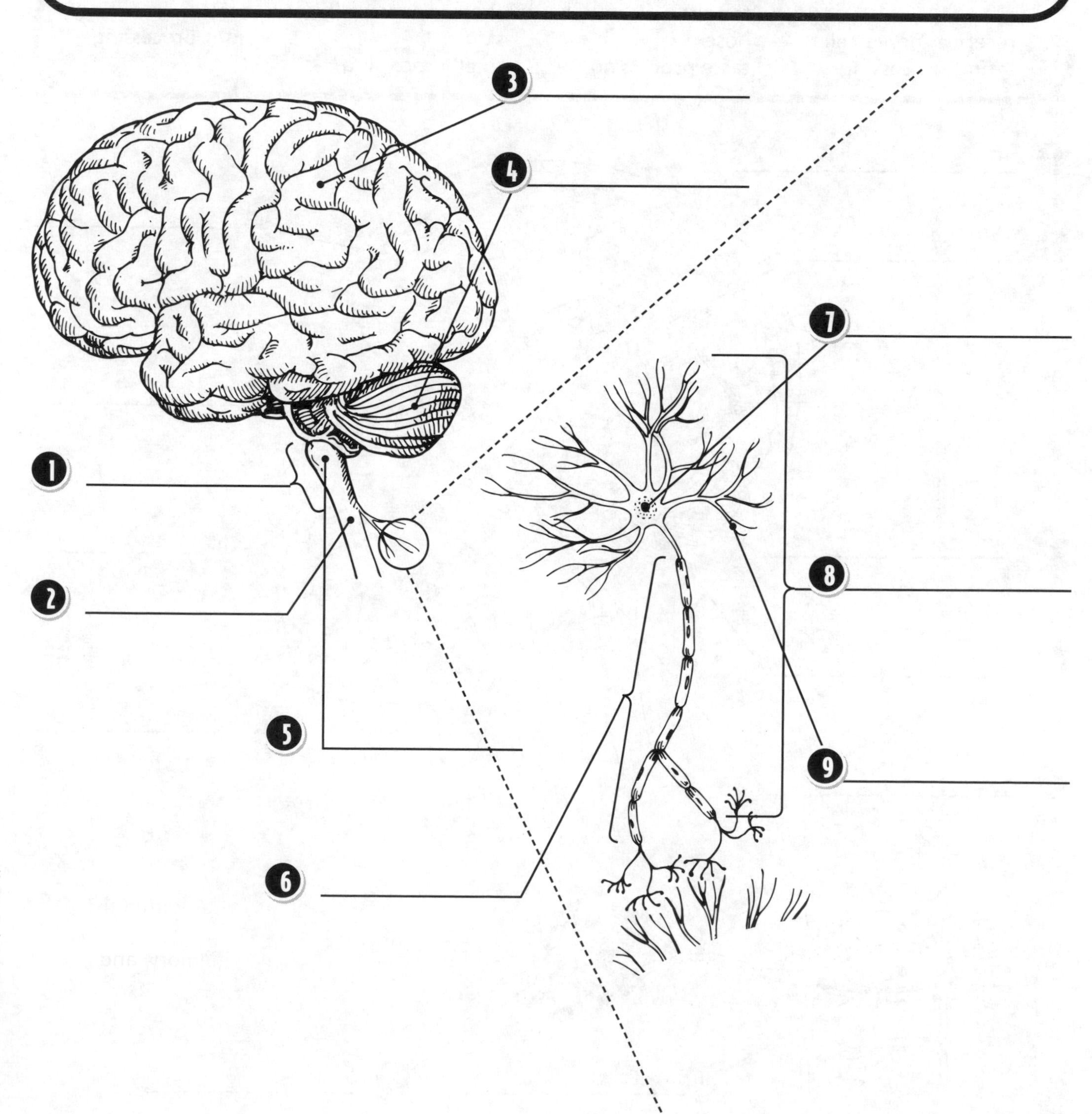

Name ______________________________ Date ______________

Sensory Systems

The brain receives information from outside of the body through many different sense organs. Each sensory organ sends its messages to a specific part of the brain for processing. Use the terms in the word box to label the diagram.

eye	tongue	sensory nerve cell	ear
receptor nerve cell	nose	skin	sight processing
sound processing	taste processing	smell processing	

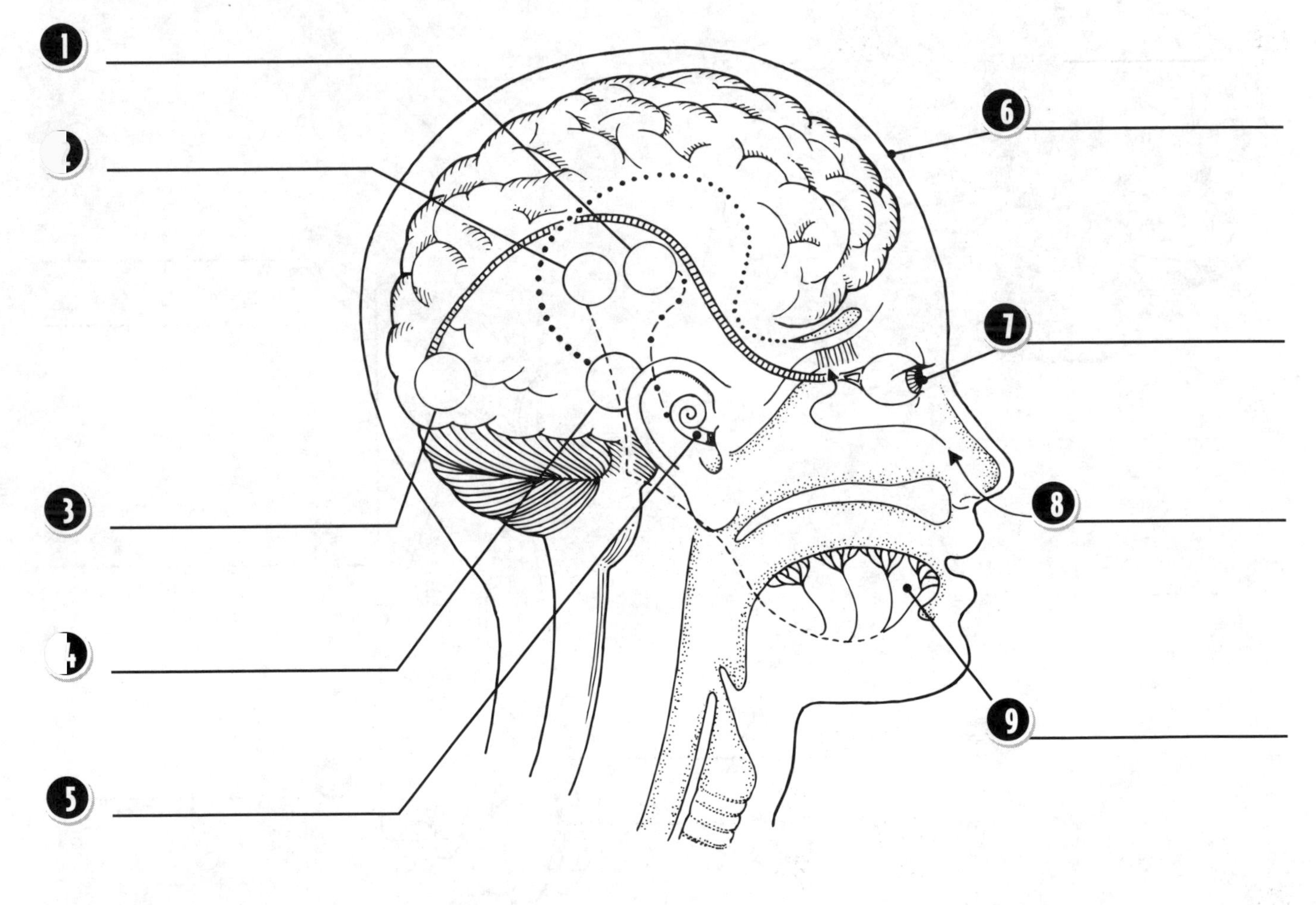

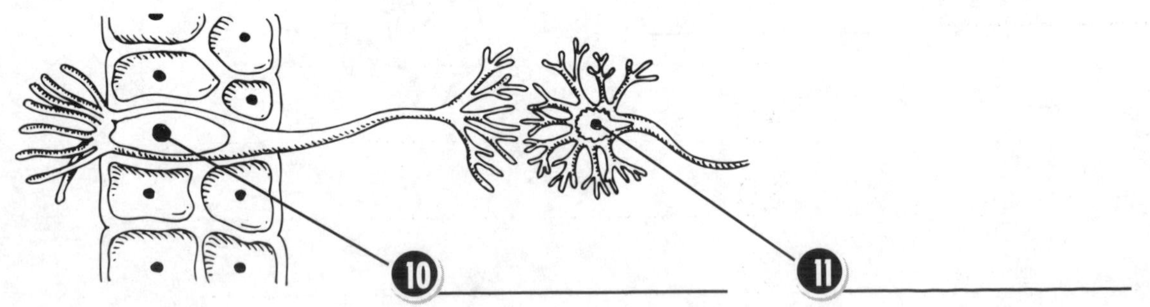

Name ______________________________ Date ______________

The Ear

The ear is an organ not only of hearing but of balance as well. The ear consists of three sections: the inner ear, the middle ear, and the outer ear. The outer and middle ear function only in hearing, while the inner ear also controls balance and orientation. Use the terms in the word box to label the diagram.

auditory canal	hammer	stirrup	semicircular canals
anvil	auditory nerve	oval window	eardrum
Eustachian tube	wax gland	auricle	cochlea

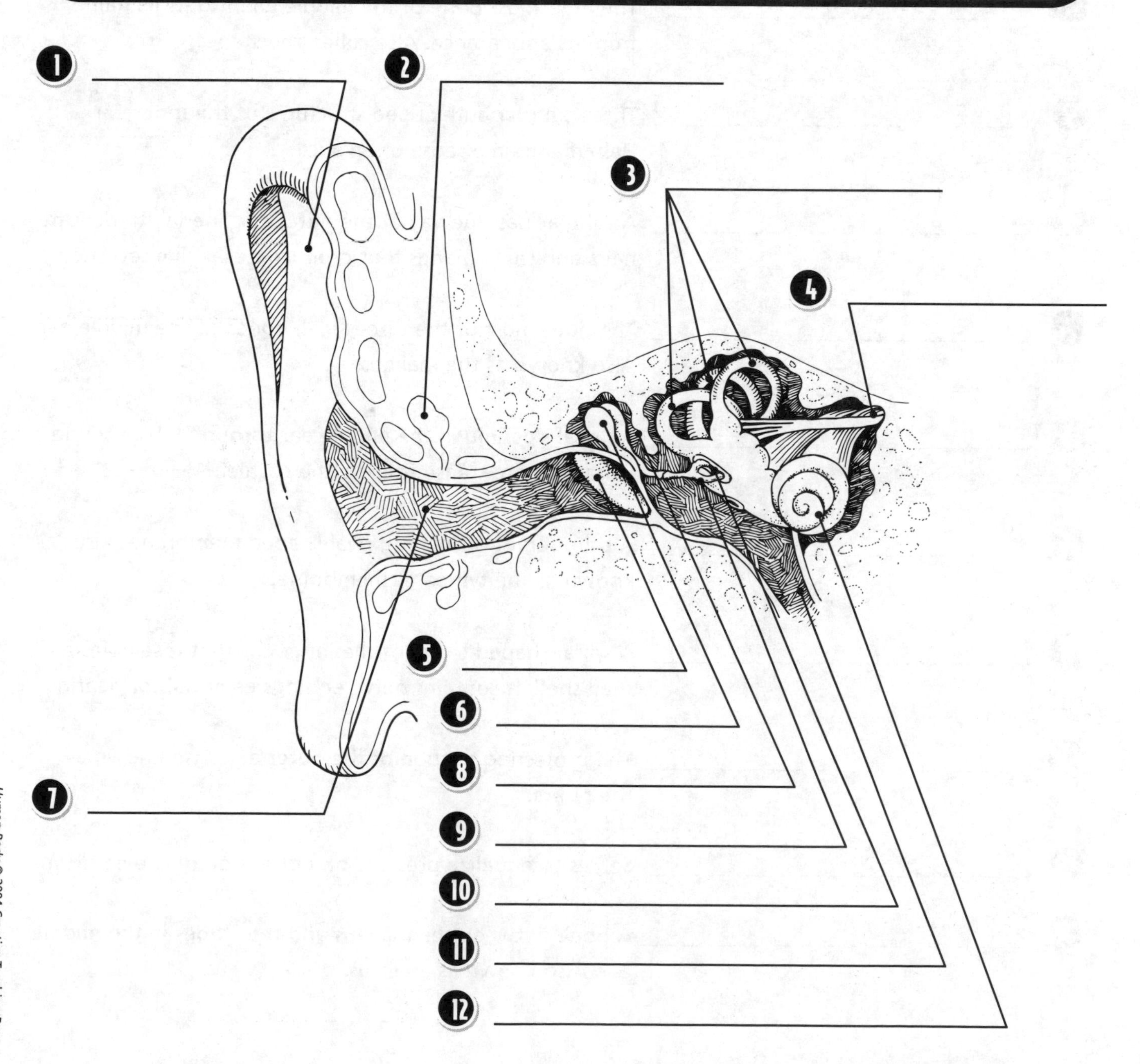

Name ______________________ Date ____________

Functions of the Ear Parts

Each part of the ear serves a specialized function for hearing or balance. Match each term in the word box to its description.

auditory canal	hammer	stirrup	semicircular canals	anvil
oval window	eardrum	auricle	Eustachian tube	cochlea

1. ______________________ The innermost bone of the middle ear, it gets its name from its appearance. Also called the stapes.
2. ______________________ Three tubular and looped structures of the inner ear. Help maintain a sense of balance.
3. ______________________ A tubular passageway of the outer ear lined with delicate hairs and small glands that produce a wax-like secretion.
4. ______________________ The outermost of the three small bones of the middle ear. Also known as the malleus.
5. ______________________ An oval opening in the middle ear through which sound vibrations are transmitted to the cochlea.
6. ______________________ A thin, semi-transparent, oval-shaped membrane. Also known as the tympanic membrane.
7. ______________________ A spiral-shaped cavity of the inner ear that resembles a snail shell. It contains nerve endings essential for hearing.
8. ______________________ The projecting portion of the outer ear. Also known as the pinna.
9. ______________________ Serves to equalize pressure on either side of the eardrum.
10. ______________________ A bone between the malleus and the stapes in the middle ear. Also known as an incus.

Name ______________________ Date __________

The Outer Ear

Functioning strictly for hearing, the outer ear is designed to channel sound waves from outside the body through the auditory canal. Use the terms in the word box to label the diagram.

outer ear	middle ear	inner ear	wax gland	auditory canal
cerumen	hairs	auricle	temporal bone	tympanic membrane

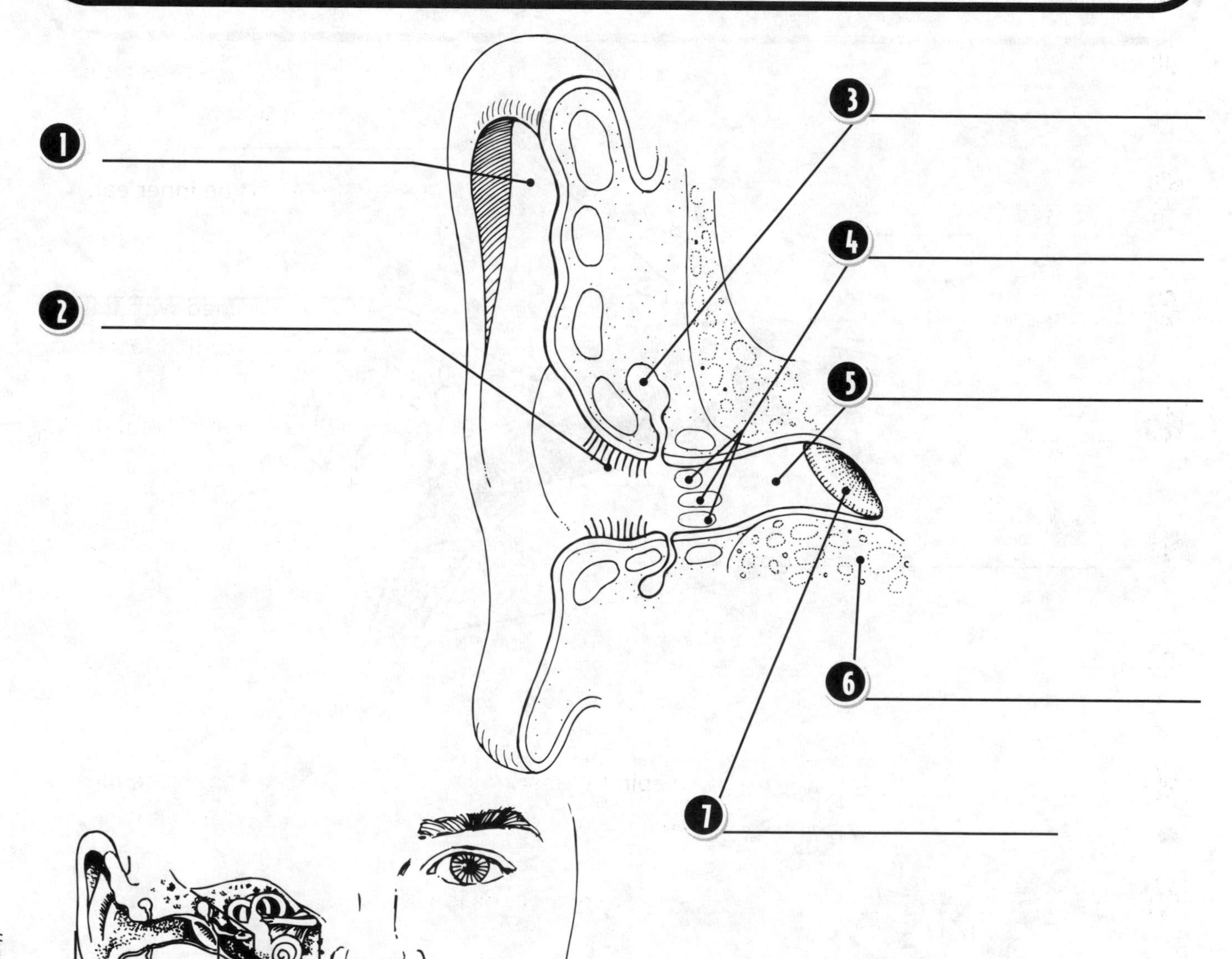

Name ______________________________ Date ______________

The Middle Ear

The eardrum separates the middle ear from the outer ear. It is in this midsection that pressure on both sides of the eardrum is regulated, causing your ears to "pop" when you travel to high altitudes. Use the terms in the word box to label the diagram.

middle ear	outer ear	inner ear	malleus
stapes	incus	Eustachian tube	

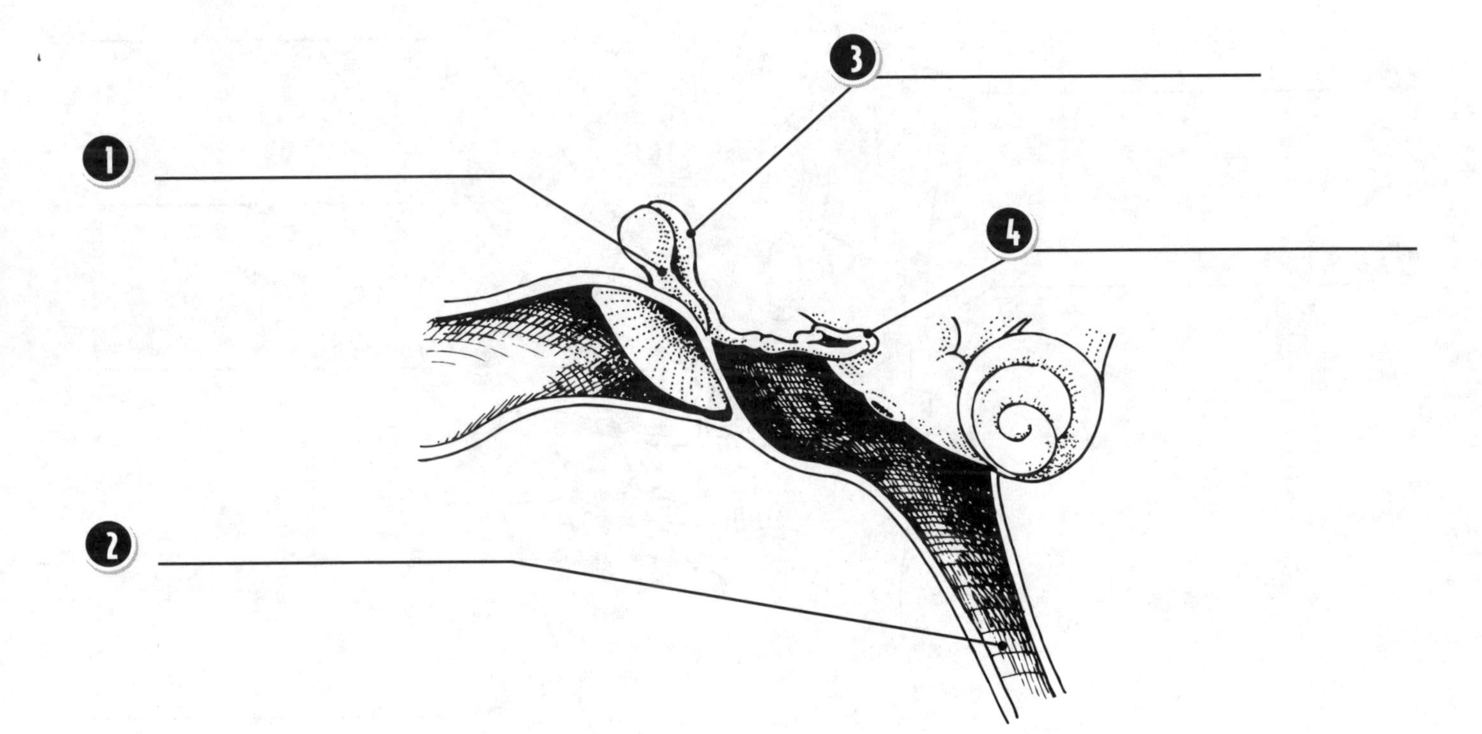

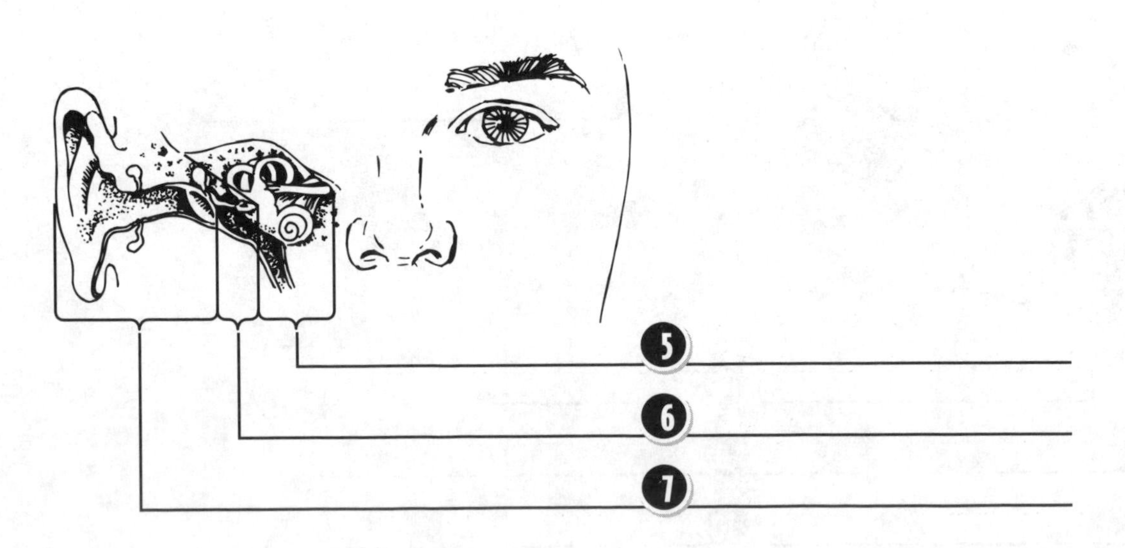

Name ______________________ Date ____________

The Inner Ear

The inner ear, also known as the labyrinth, is the innermost portion of the ear. Located entirely within the temporal bones, it contains structures for hearing and balance. Use the terms in the word box to label the diagram.

inner ear	outer ear	middle ear	cochlea
oval window	auditory nerve	semicircular canals	

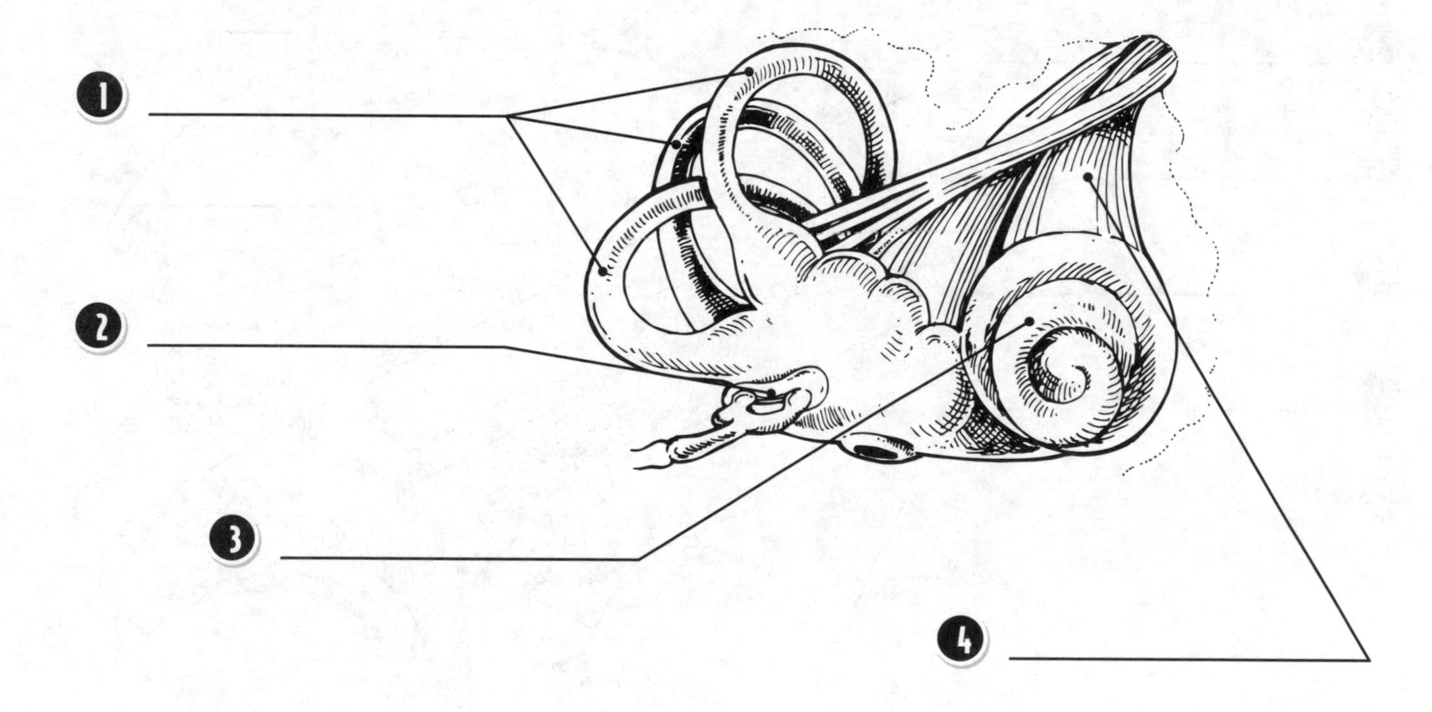

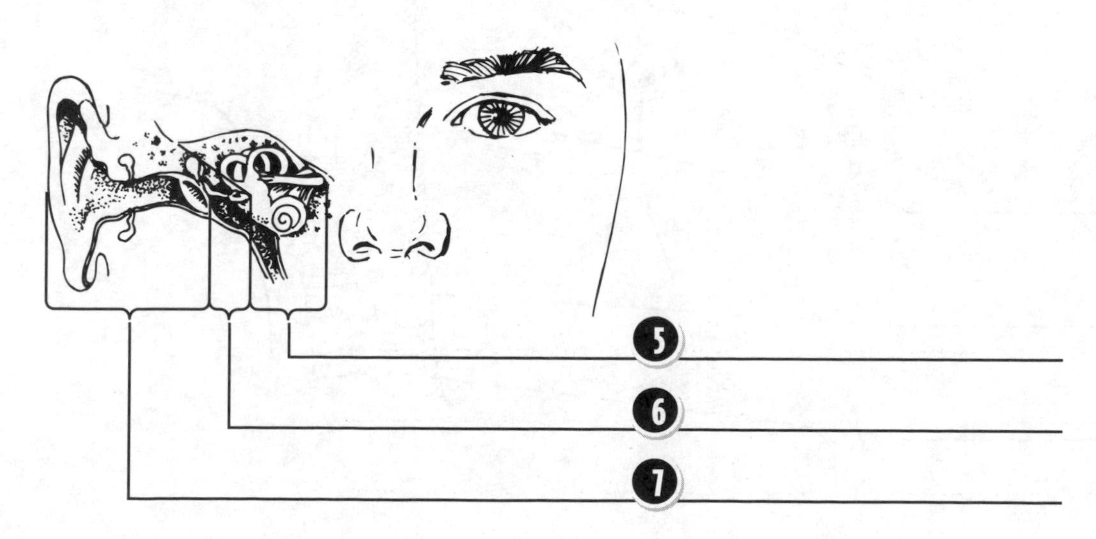

Name ______________________________ Date ______________

What's the Connection?

The ears, nose, and throat are all connected to each other. Because of these connections, the pressure behind the eardrum can be balanced. However, sometimes an infection in one part will result in symptoms or pain in another as well. Use the terms in the word box to label the diagram.

trachea	epiglottis	esophagus	Eustachian tube
nasal passage	palate	pinna	inner ear

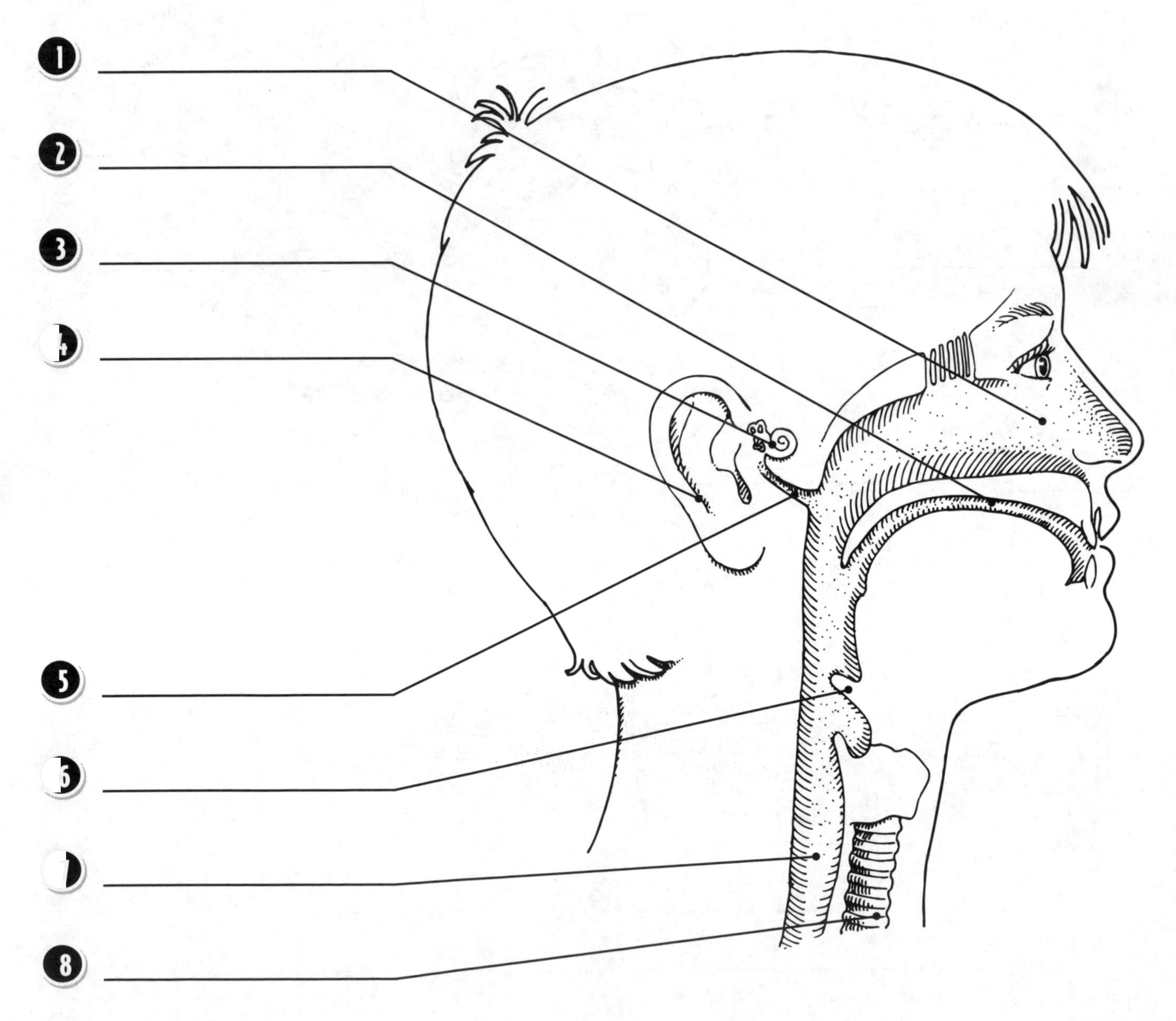

Name ______________________ Date ____________

The Eye

The human eye is a spherical structure with a pronounced bulge on its forward surface. Its function is to collect light rays and transmit them to the brain. Use the terms in the word box to label the diagram. Some words are used twice.

optic nerve	iris	sclera	retina	lens
pupil	cornea	optic disk	vitreous body	aqueous humor

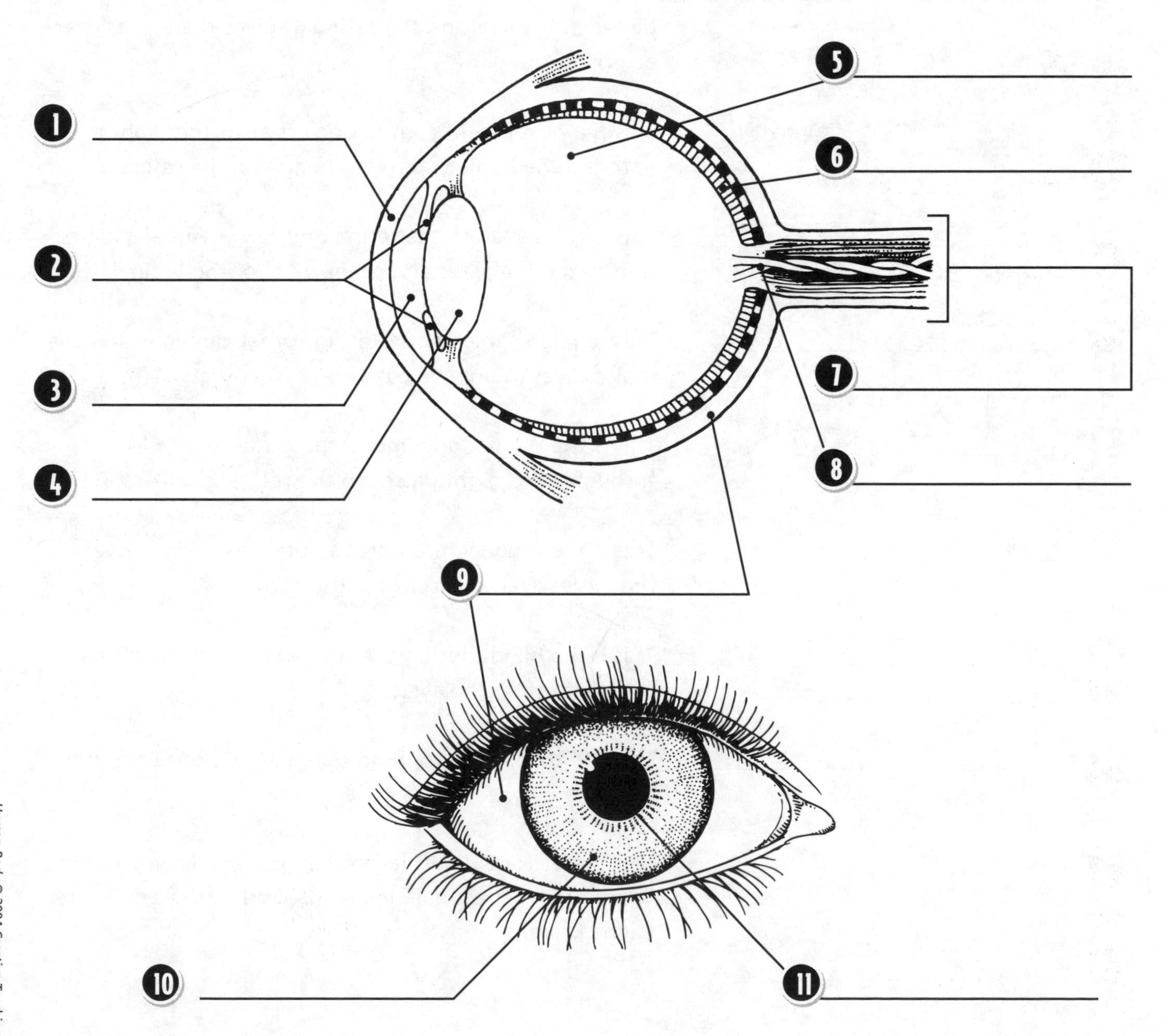

Name ______________________________ Date ______________

Parts of the Eye

The parts of the eye work together to collect light rays, transmit them into electrical impulses, and send these impulses to the brain. Match each term in the word box to its description.

optic nerve	retina	lens	pupil	iris
sclera	optic disk	aqueous humor	vitreous body	

1. ______________________ This is the clear gelatinous substance that fills the eyeball between the lens and the retina. It gives the eye its characteristic roundness.

2. ______________________ This is a transparent convex body that focuses light rays entering the pupil to form an image on the retina.

3. ______________________ This is connected to the retina and carries visual information to the thalamus and other parts of the brain.

4. ______________________ This regulates the amount of light that can enter the eye. It also gives us our eye color.

5. ______________________ This is the circular opening in the center of the iris through which light passes to the retina.

6. ______________________ This is the tough white fibrous outer tissue that covers the entire eyeball except for the cornea.

7. ______________________ This is the blind spot on the retina where the optic nerve originates from the eye.

8. ______________________ This is the clear watery fluid between the cornea and the lens of the eye.

9. ______________________ This is a delicate, multi-layered, light-sensitive membrane lining the inner eyeball and connected by the optic nerve to the brain.

Name ______________________________ Date ______________

Inside the Eye

Looking more closely at the inner parts of the human eye, there are many specialized parts that help us to see. Use the terms in the word box to label the diagram.

optic nerve	choroid	sclera	retina	lens
pupil	cornea	optic disk	vitreous body	aqueous humor
iris	ciliary muscles	rod cells	cone cells	

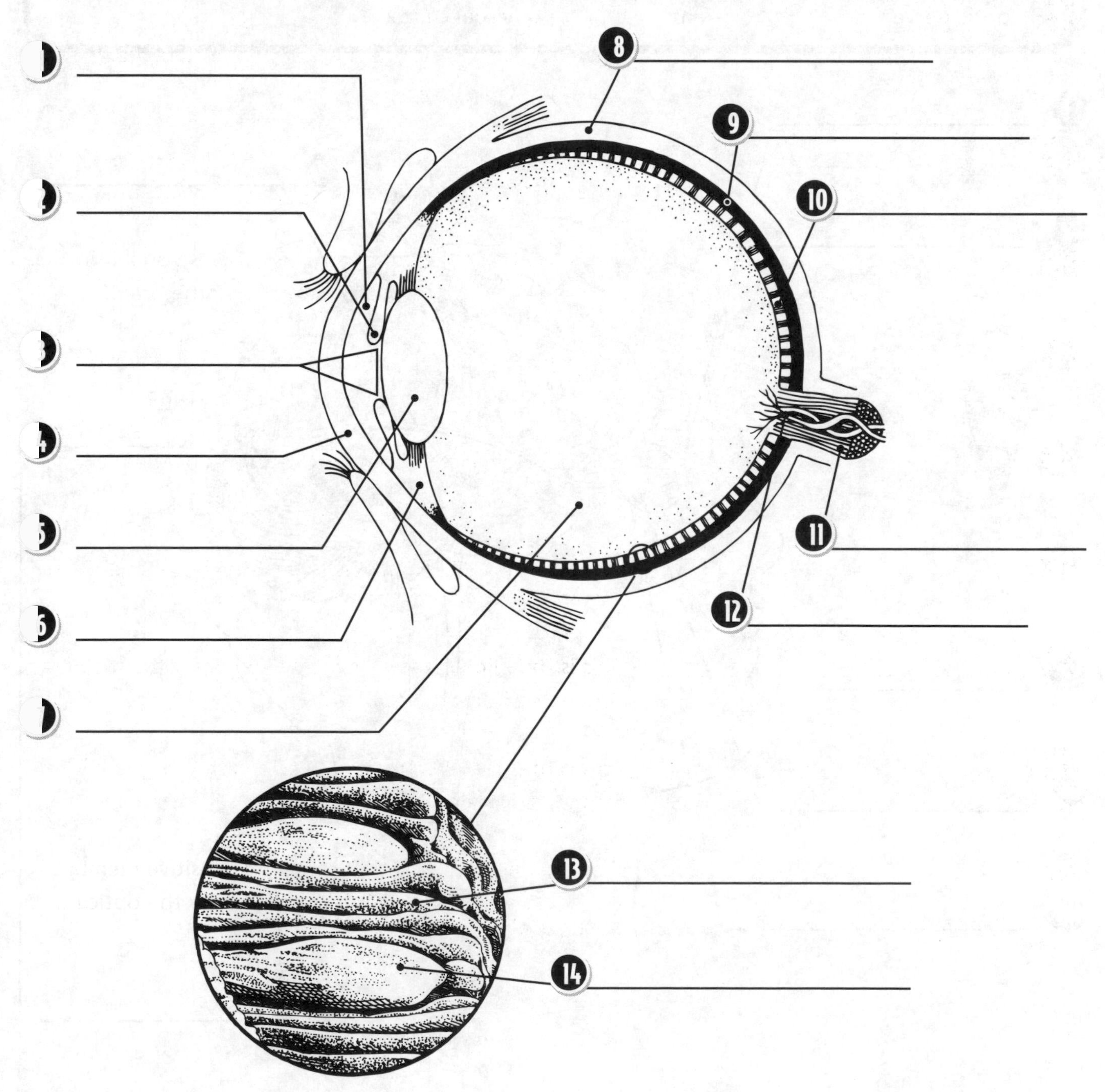

Name ______________________________ Date ______________

I See, I Understand

The human eye collects light rays that bounce off of objects. As the image passes through the lens, it is turned upside down. Light rays that reflect from the upper half of any object we look at are focused on the lower half of the retina. Rays from the lower half of the same object are focused on the upper half of the retina. These signals are transmitted to the brain through the optic nerve. In the brain, they are rearranged into an image that is right side up. Use the terms in the word box to label the diagram.

image of object	retina	upside-down image	cornea
optic nerve	lens	visual cortex	

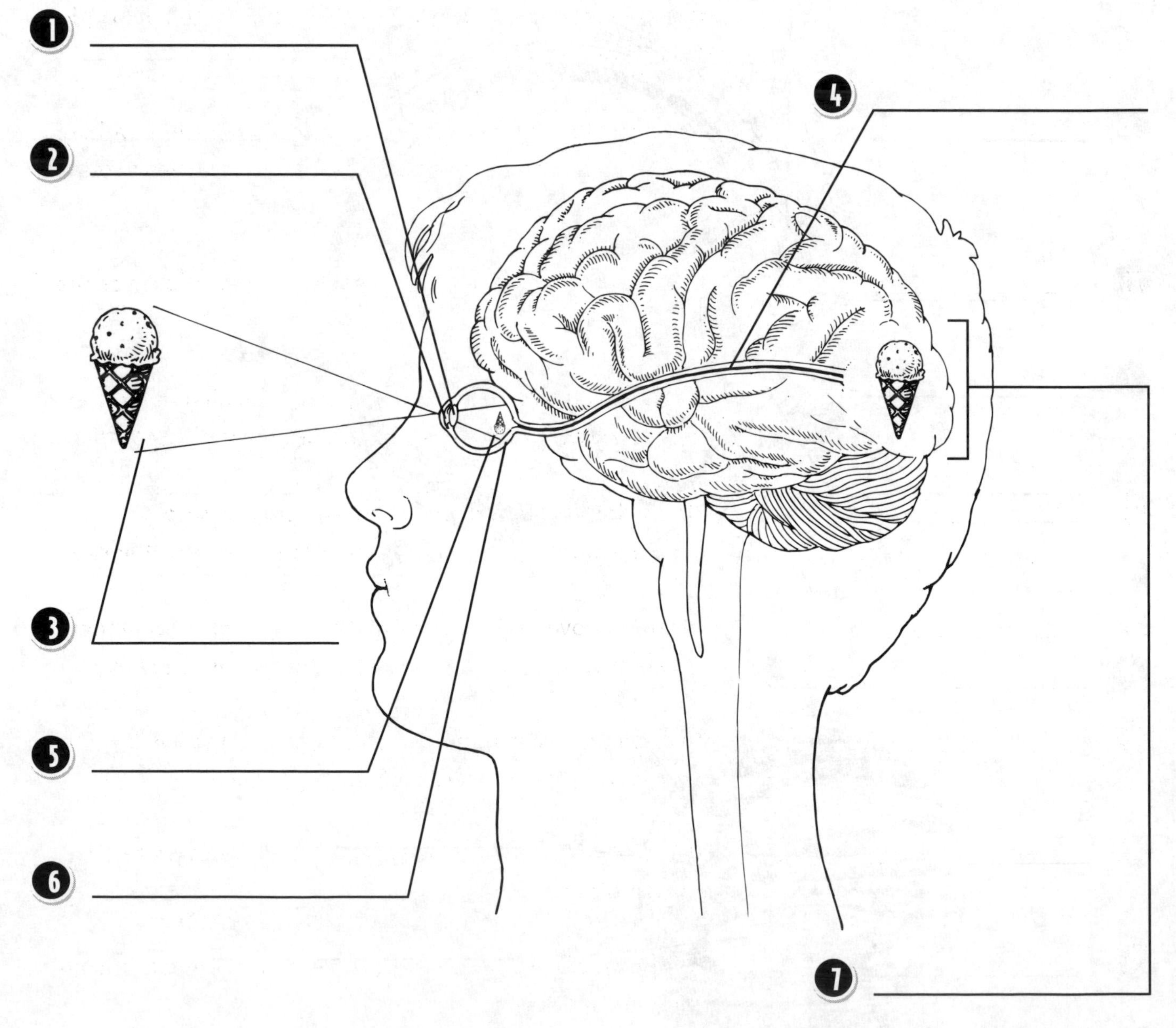

Name ______________________________ Date ______________

The Protection Connection

Several structures and their functions contribute to the protection of the eye. Match each term in the word box to its description.

eyelashes	blinking	conjunctiva	eyelids	tear duct
skull	eyebrows	frontal bone	cheekbone	

1. ______________ These upper and lower folds of skin and tissue can be closed by means of muscles. They form a protective covering over the eye that protects it from receiving too much light and from injury.
2. ______________ A fringe of short hairs growing on the edge of the eyelids. They act as a screen to keep dust particles and insects out of the eyes.
3. ______________ A thin protective membrane that doubles over to cover the visible sclera.
4. ______________ Situated in the corner of the eye, this small gland produces a salty secretion that lubricates and flushes away tiny particles.
5. ______________ This action happens by reflex about every six seconds. It happens more often if dust reaches the surface of the eye.
6. ______________ Located above each eye, they protect by keeping rain, perspiration, and other moisture from running into the eyes.
7. ______________ The hollow socket in this structure protects the eye by surrounding it with sturdy bone.
8. ______________ This protrusion of bone above the eye provides protection against blows or collisions.
9. ______________ This protrusion of bone below the eye also protects it from injury from blows or collisions.

Name ______________________ Date ____________

Protecting the Eye

Use the terms in the word box to label the diagram.

pupil	iris	eyelashes	sclera	eyelids
tear duct	skull	eyebrows	frontal bone	cheekbone

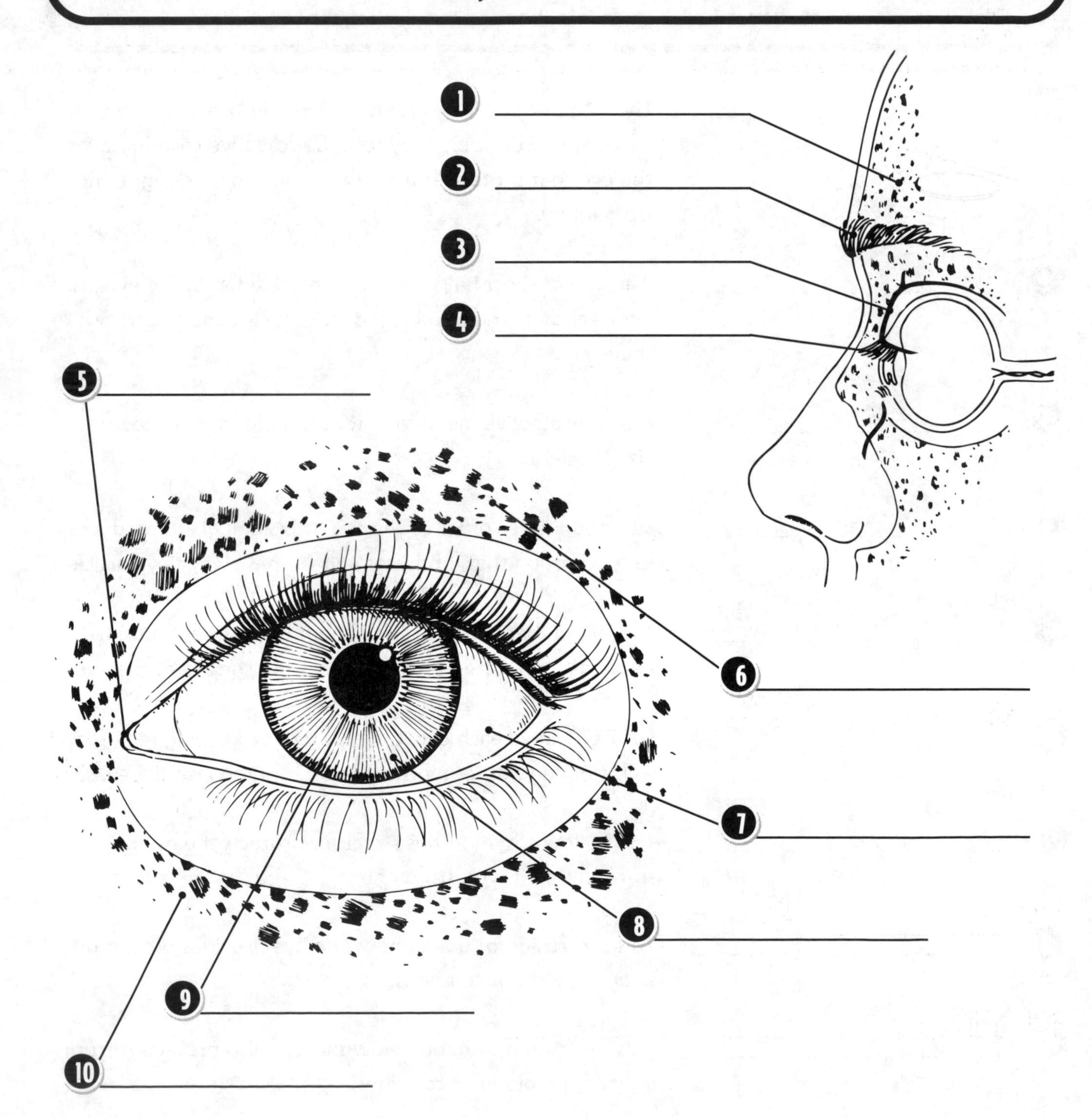

Name ______________________________ Date ______________

Just Like a Camera

The human eye operates in a manner very similar to a camera. In fact, the parts of a camera have their equivalent in the eye. Use the terms in the word box to label each diagram. Some words are used twice. Use the phrases to describe the functions on the line below the labels.

retina	iris	lens	film	used to focus
adjusts the amount of entering light		light sensitive material that records an image		

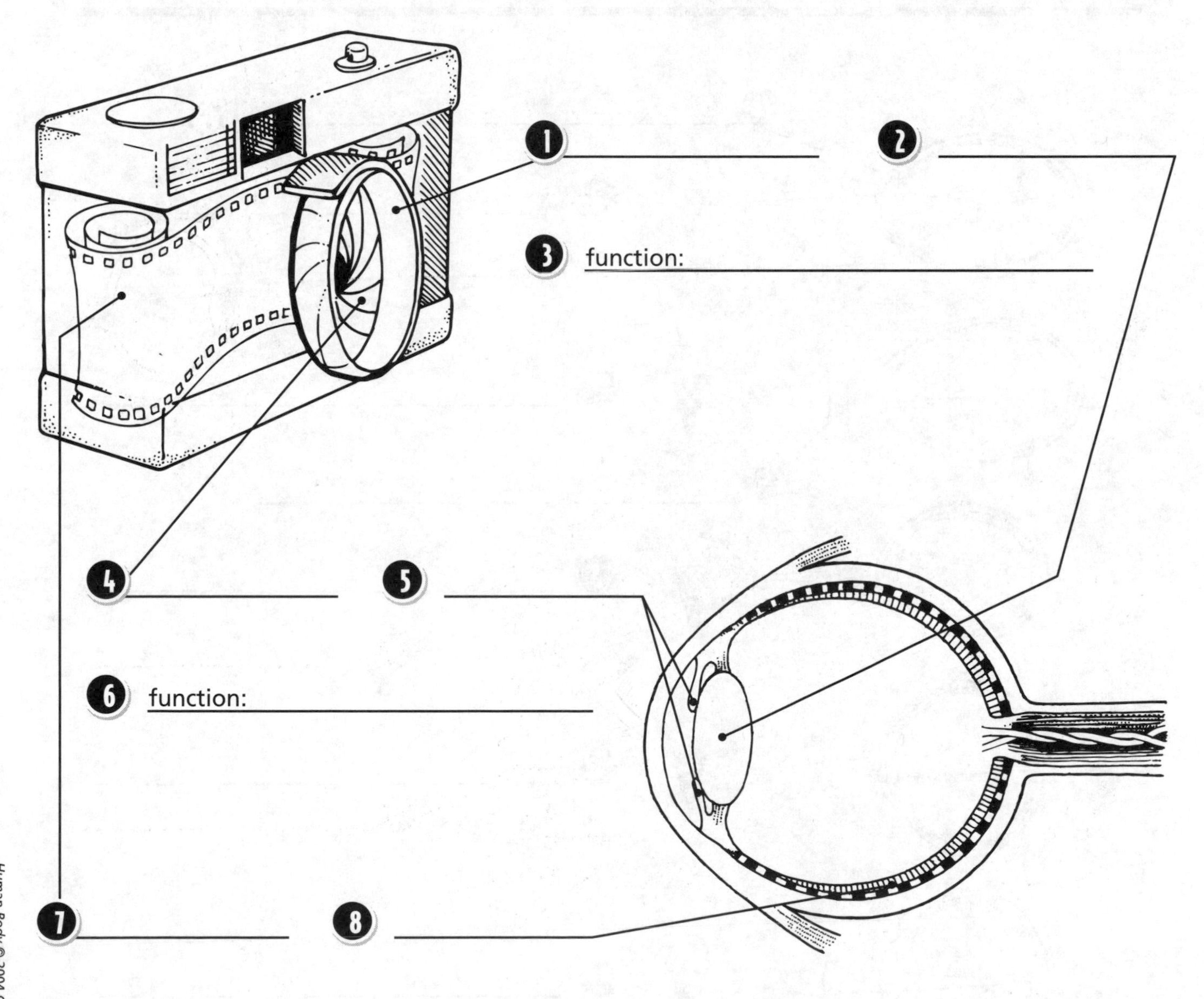

9 function: ______________________________

Name ______________________ Date ______________

Eyesight and Shape

The eyeball can vary in shape. Sometimes the shape affects how well the eye can focus an image. In those cases, corrective lenses are needed to help position the image the right distance within the eye to help it see clearly. Use the terms in the word box to label each diagram.

normal vision	farsighted vision	corrected farsighted vision
concave lens	lens	nearsighted vision
convex lens		corrected nearsighted vision

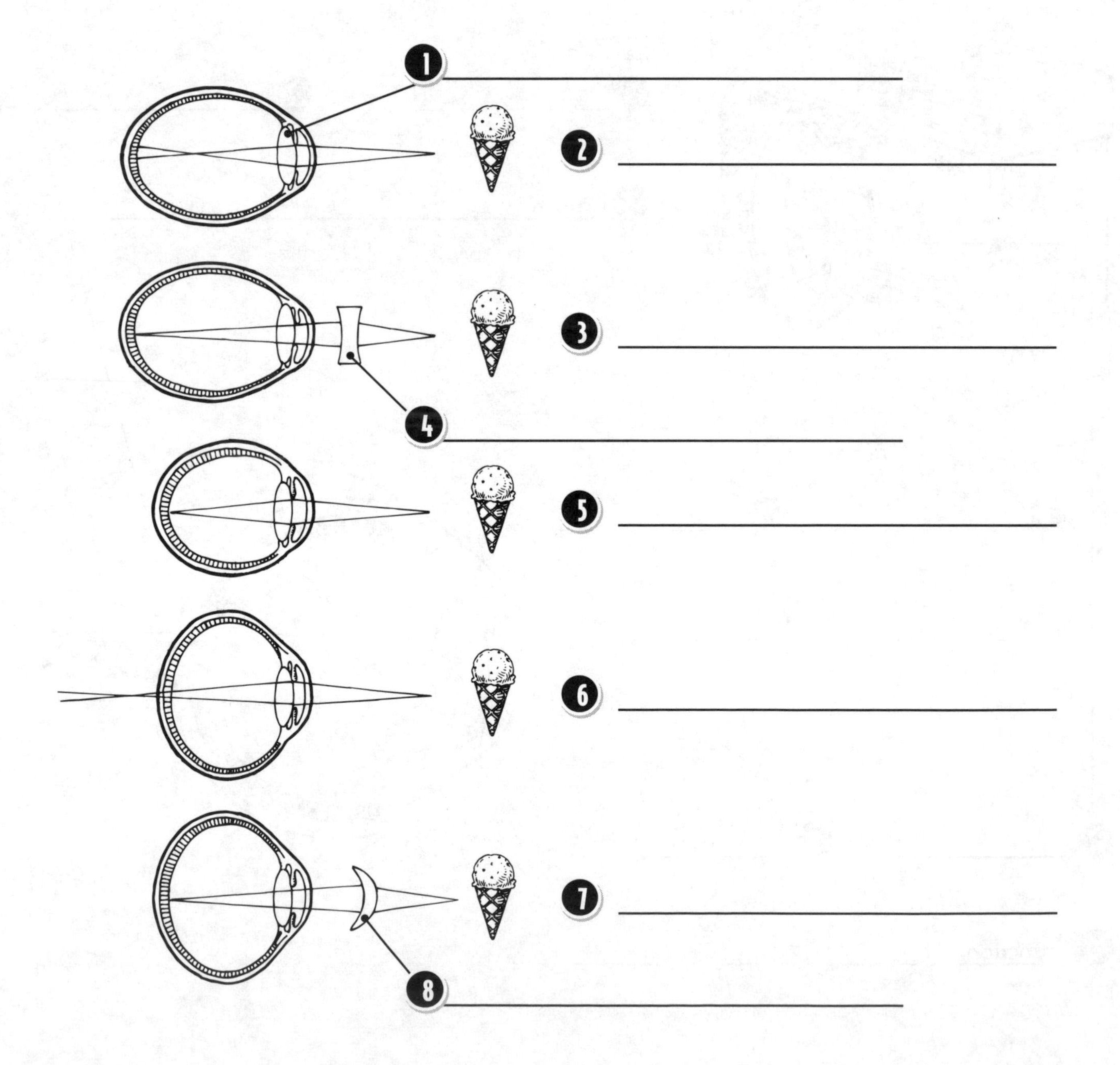

Name ______________________________ Date ______________

Ear and Eye Review

Review what you have learned about the parts of the eye and ear. Use the terms in the word box to label each diagram.

cornea	iris	auditory nerve	lens	sclera
malleus	incus	stapes	auricle	retina
pupil	optic nerve	cochlea	eardrum	semicircular canals

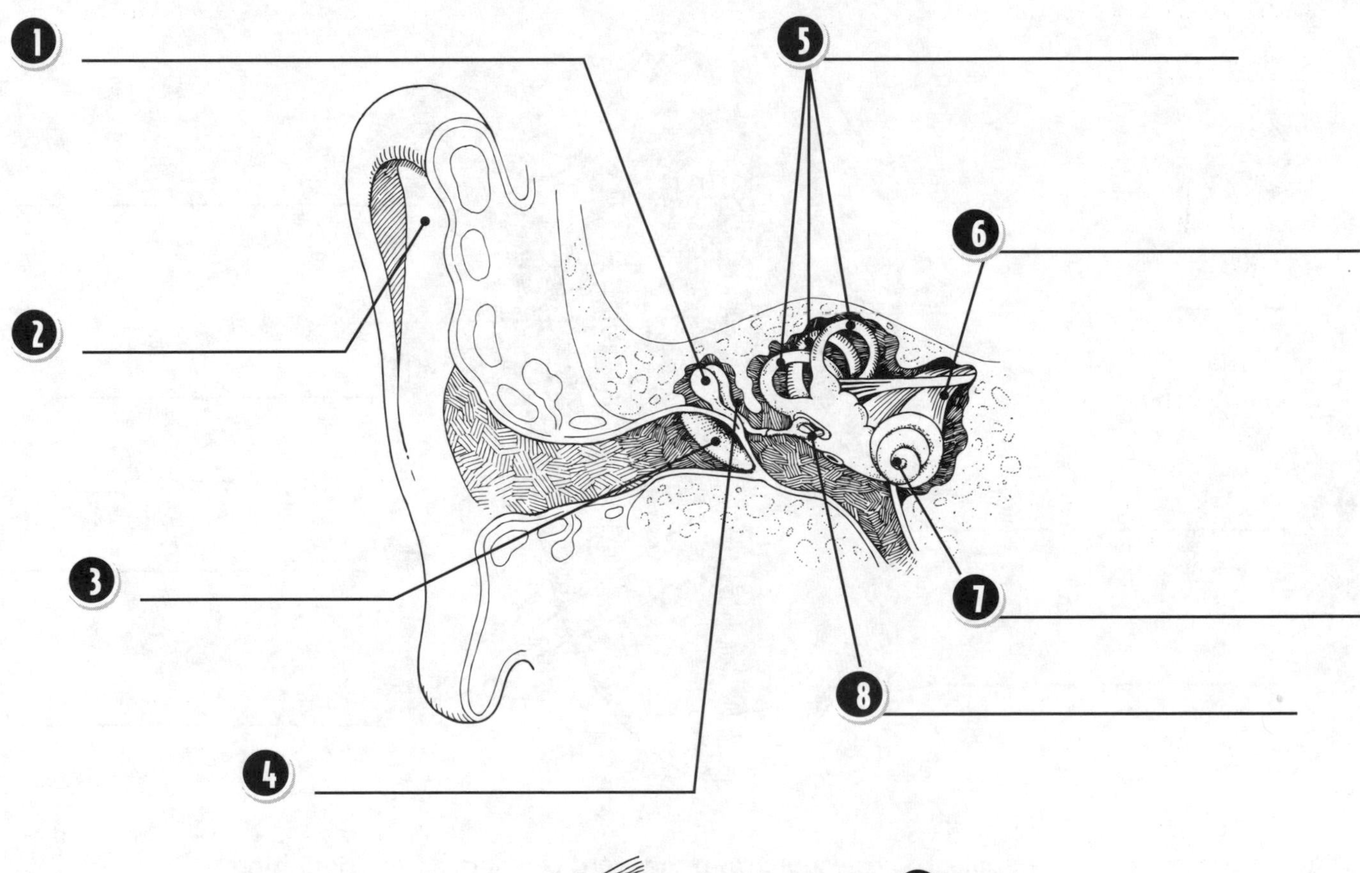

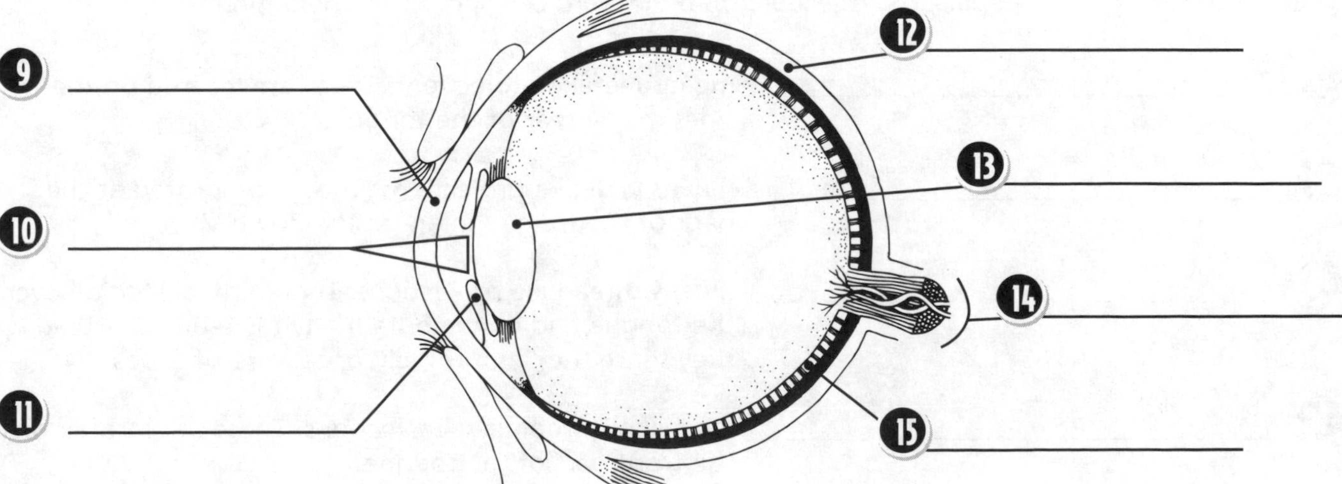

Name ______________________ Date ____________

The Organ of Taste

Taste is the ability to sense dissolved chemicals in materials placed in the mouth. The tongue is the organ that distinguishes taste. Its surface is covered with papillae, or projections. Use the terms in the word box to label the diagram.

palatine tonsil	lingual tonsil	foliate papillae
filiform papillae	fungiform papillae	circumvallate papillae

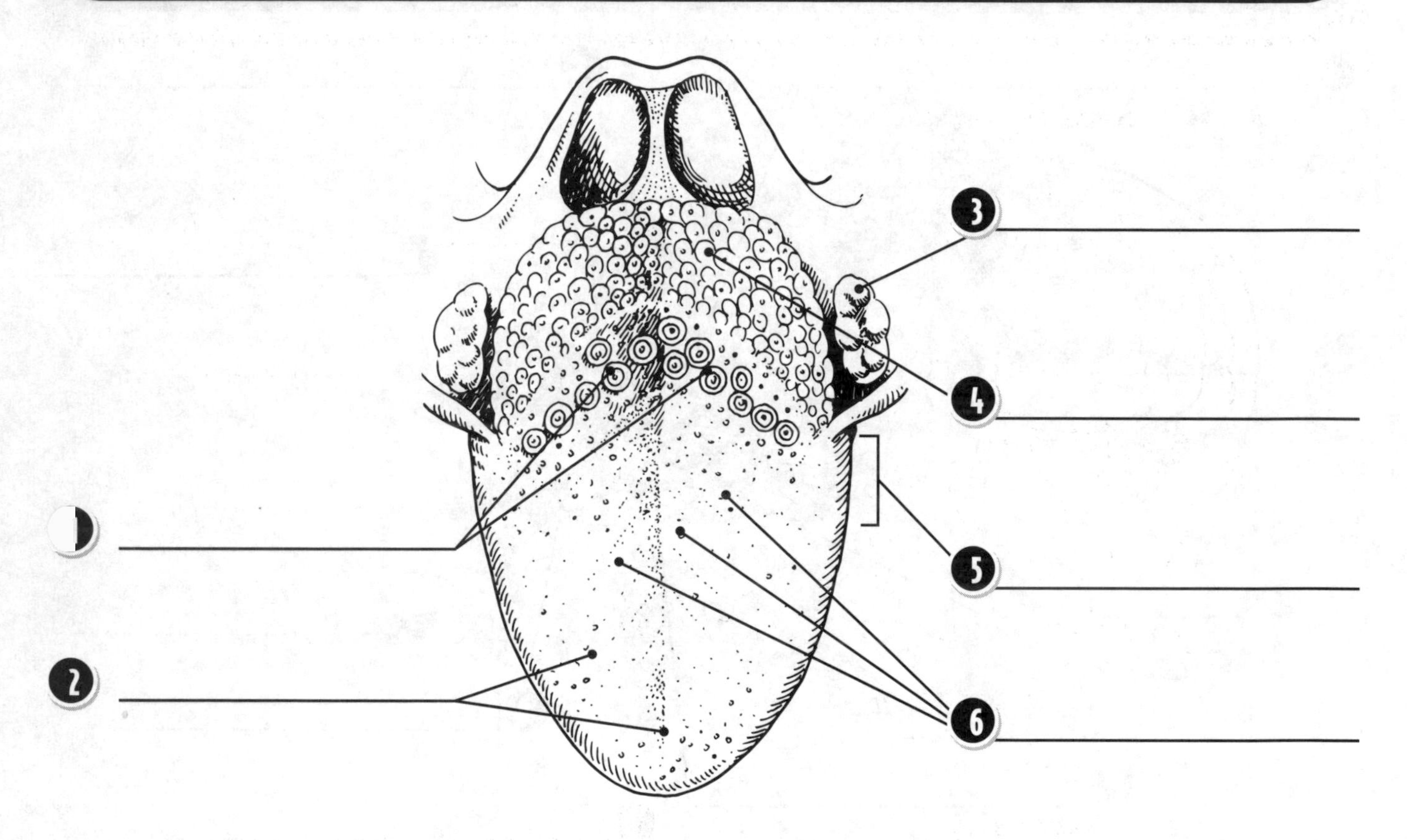

Match the name of each papilla, or projection, from the word box above to its definition.

7. ______________________ means leaflike projections. These are located on the sides of the rear of the tongue.

8. ______________________ means wall-like projections. These are located at the back of the tongue in an upside-down V.

9. ______________________ means threadlike projections. These are located all over the tongue and don't contain any taste buds. Instead, they sense the "mouthfeel" or texture of foods.

10. ______________________ means mushroomlike projections. These are located on the front part of the tongue.

Name ______________________________ Date ______________

Taste Buds

Within the papillae of the tongue are the taste buds. Taste buds are onion-shaped structures. Each one contains between 50 and 100 taste cells. Chemicals from foods dissolve in the saliva and come in contact with the taste cells. The types of chemicals the cells pick up determine the flavors we taste. Use the terms in the word box to label the diagrams.

nerve fiber	brain	epithelium	taste pore	tongue
taste cell	taste bud	microvilli	nerve	connective tissue

Sensing Taste

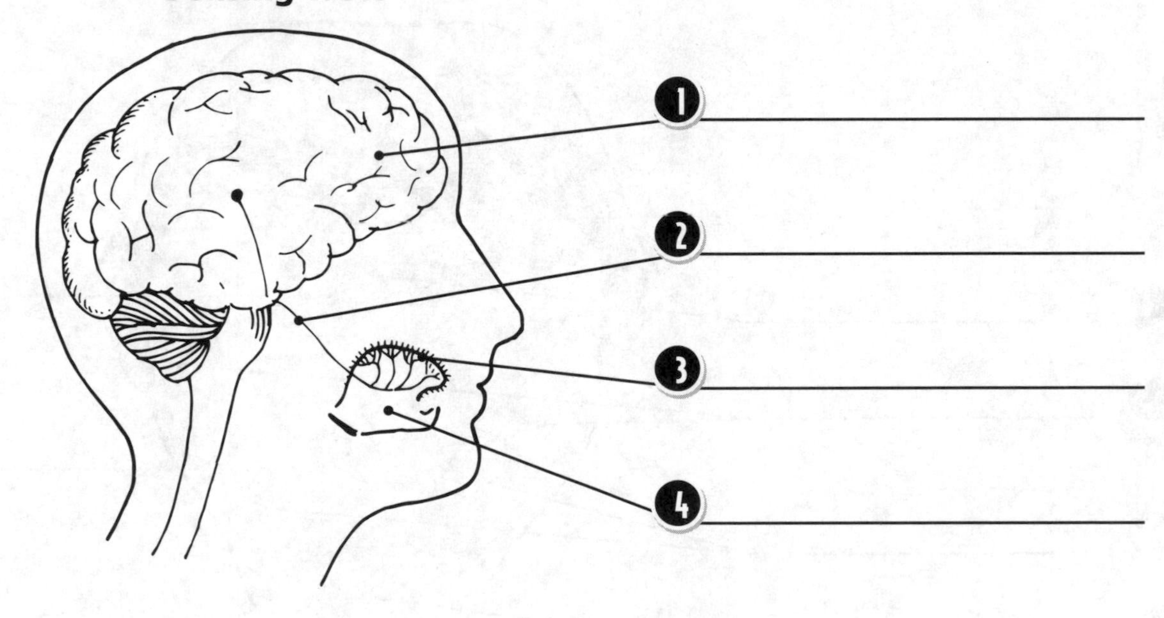

1. ______________________
2. ______________________
3. ______________________
4. ______________________

Taste Bud

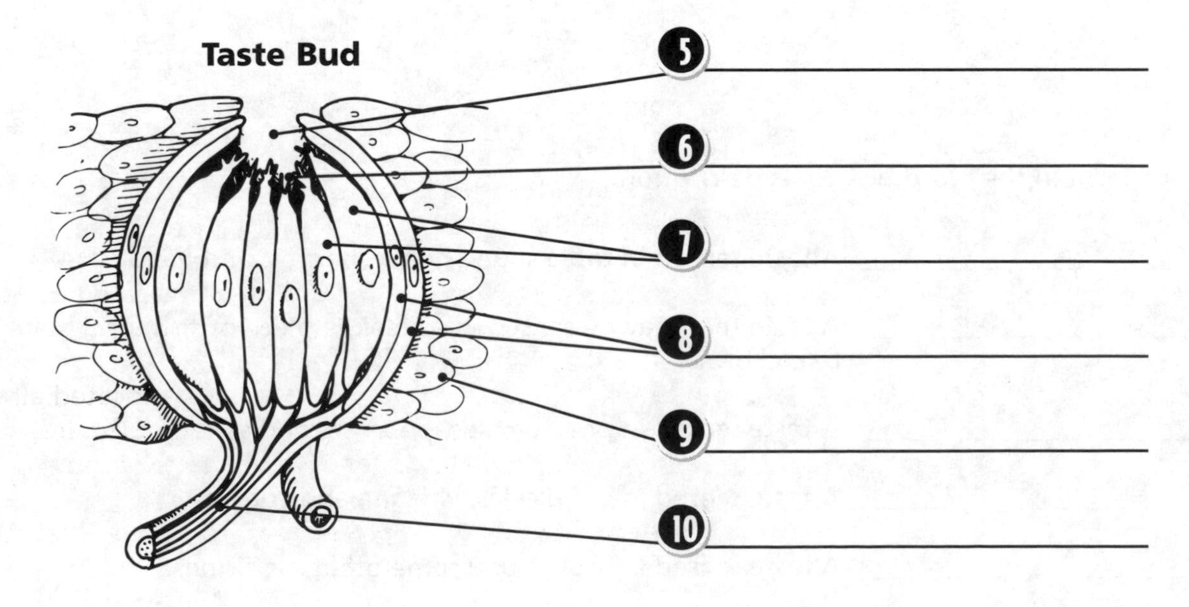

5. ______________________
6. ______________________
7. ______________________
8. ______________________
9. ______________________
10. ______________________

Name ______________________ Date ______________

How Does That Taste?

Taste is described by four qualities: sweetness, saltiness, sourness, and bitterness. Some scientists also think there is a fifth taste, umami. This describes the taste of amino acids present in the proteins of meat, fish, legumes, and monosodium glutamate (MSG). While all taste cells can distinguish all tastes, some seem more receptive to certain flavors than others. Use the terms in the word box to label the diagram.

salty	sweet	sour	bitter	umami

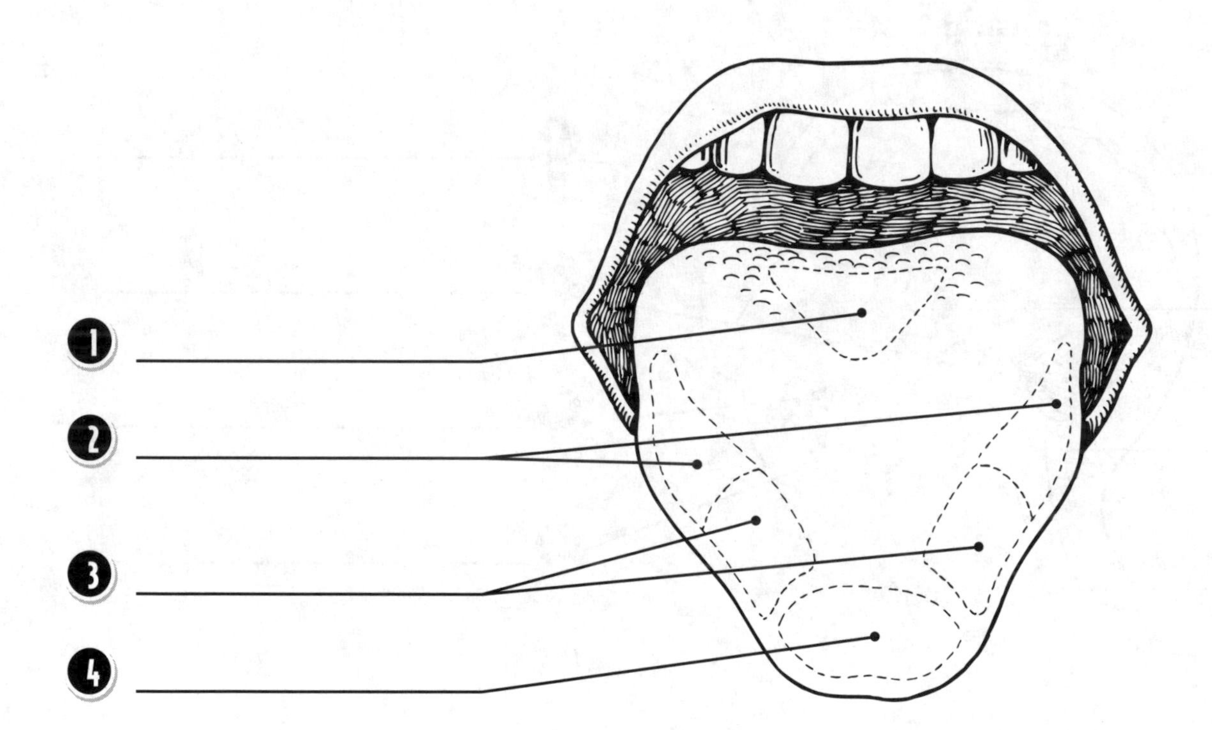

Match each term in the word box to its description.

5. ______________ A taste sensed in table sugar, honey, berries, and melons.
6. ______________ A taste that may be sensed in a steak, a piece of fish, or a bowl of baked beans.
7. ______________ A taste sensed in potato chips, popcorn, and salted nuts.
8. ______________ A taste sensed in dill pickles, vinegar, and yogurt.
9. ______________ A taste sensed in aspirin and some other medicines.

Name ______________________ Date ____________

The Nose and Smell

Smell is the sense by which odors are perceived. The nose is the special organ of smell, although it also has a role in respiration and voice. The nose most often refers to the external portion, the part we see that protrudes from our faces. However, the nose leads to internal portions that can be divided into two main cavities separated by the septum. Use the terms in the word box to label the diagram.

nostril	olfactory nerve	brain	nasal passage
receptor cells	tonsil	nasopharynx	tongue
sphenoidal sinus	frontal sinus	cartilage	teeth

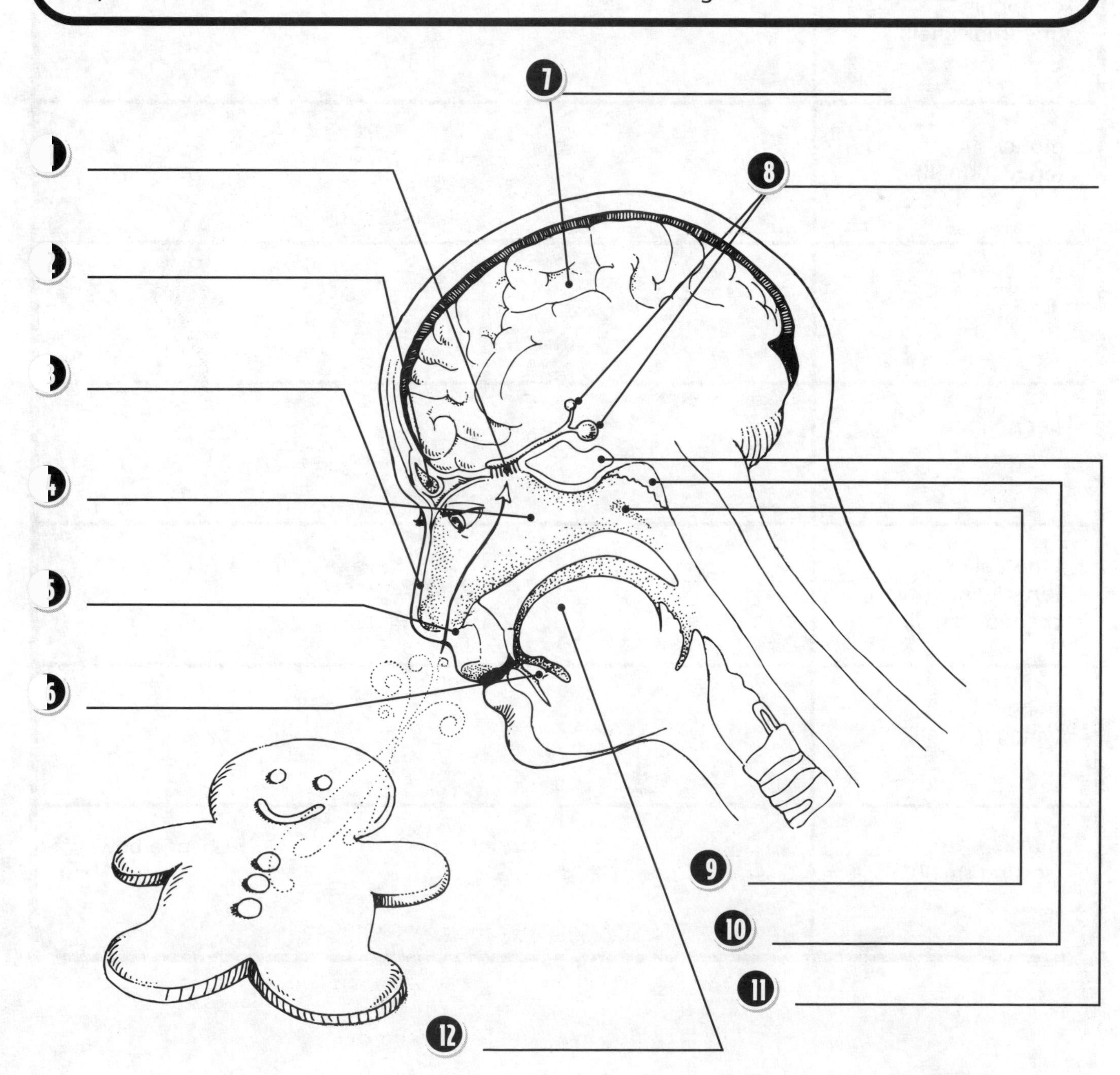

Name ______________________________ Date ______________

Making Sense of Smell

Sensations of smell are difficult to describe and classify. It seems that everything has its own unique smell, if it smells at all. Research has determined seven primary odors: camphor, musky, floral, peppermint, ethereal, pungent, and putrid. Smell is also closely linked to memory; a certain scent will often trigger a particular memory. List your own examples of each primary scent to complete the chart.

Scent	Objects of That Scent
camphor (menthol smell)	
musky (animal smell)	
floral	
peppermint	
ethereal (dry-cleaning fluid, chemical smell)	
pungent (vinegar smell)	
putrid (rotting smell)	

Name ______________________________ Date ______________

A Smelly Process

Use the terms in the word box to complete the sentences and discover interesting facts about our sense of smell.

chemicals	odor molecules	brain	cilia
mucus	head cold	scent	olfactory receptors

1. Our nose is able to discriminate between different ______________________ in the air.

2. Odor molecules are tiny bits of ______________________ floating in the air. They come from food, flowers, rotten garbage, or other things with smells.

3. About the size of a postage stamp, the ______________________ are microscopic nerve cells that detect smell and send the signals onto the brain.

4. Many odors are not just a single ______________________ but a whole mixture of them.

5. Far back in the nasal cavity, small hair-like structures called ______________________ sweep back and forth.

6. The cilia are covered with a thin, clear ______________________ that dissolves odor molecules not already in vapor form.

7. When the mucus becomes too thick, such as when we have a ______________________, it can no longer dissolve the molecules and we don't smell as well.

8. To identify the smell of a rose, the ______________________ has to analyze over 300 different odor molecules.

Name ______________________________ Date ______________

The Organ of Touch

The skin that covers our bodies is made up of many layers. These layers contain hairs, nerves, blood vessels, and glands. Use the terms in the word box to label the diagram.

epidermis	dermis	fat layer	hair muscle	fat cells	sweat gland
hair	oil gland	blood vessel	pore	nerve	

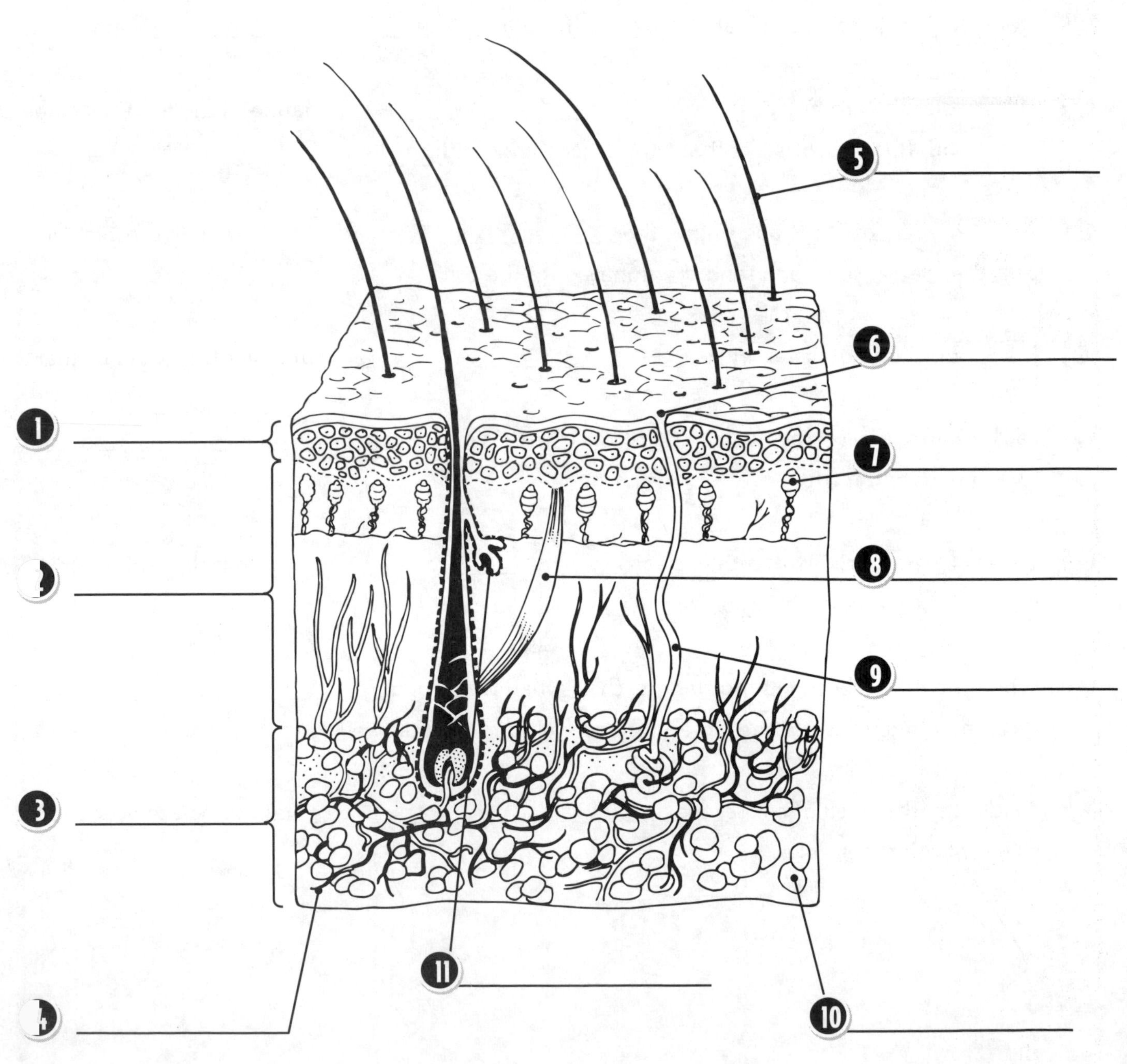

Name ______________________________ Date ______________

Blemish, Pimple, Zit

Whether you call it a blemish, a pimple, or a zit, all three are caused by the same thing: acne. Acne is an eruptive skin disorder of the sebaceous follicles of the skin. The sebum of the follicles accumulates and mixes with dust and dirt. The follicles and surrounding tissue become inflamed and blackheads appear. If the follicle opening closes, bacteria accumulate and a pimple forms. Use the terms in the word box to label the three stages shown in the diagram. Then label the diagram.

sebum gathers	follicle becomes clogged		a pimple forms
sebaceous gland blockage	hair pimple	sebum epidermis	pus hair follicle

1. ______________ 2. ______________ 3. ______________

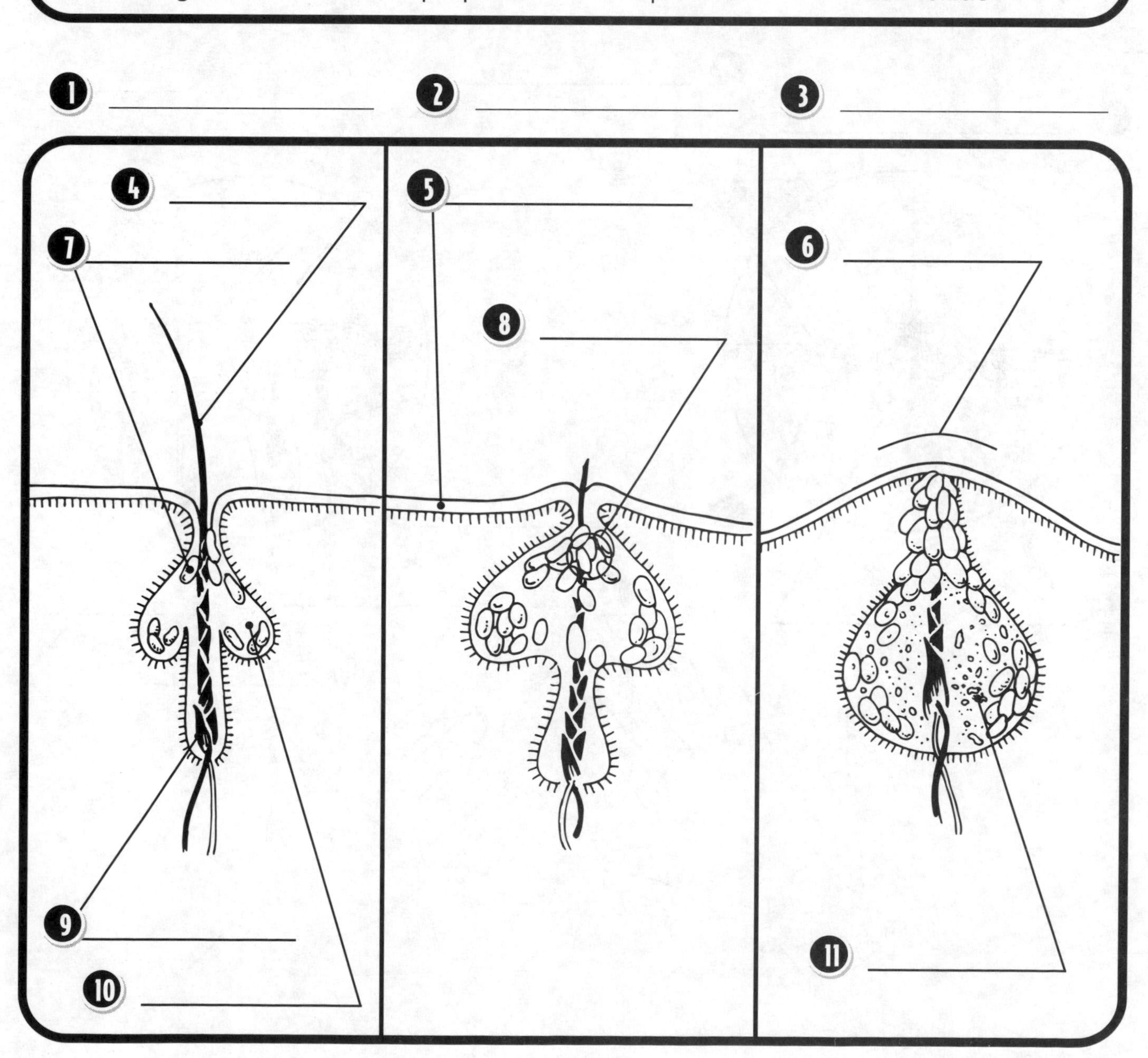

Name ______________________ Date ____________

Too Hot, Too Cold

The human body has its own way to cool off or retain warmth. When cold, the muscles around hair follicles contract to form goose bumps. When warm, the skin perspires. The moisture then evaporates, making us feel cooler. Use the terms in the word box to label the diagrams.

closed sweat pore	relaxed muscle	sweat gland	goose bump	warm day
open sweat pore	contracted muscle	blood vessel	sweat	cold day

1. ______________
2. ______________
3. ______________
4. ______________
5. ______________
6. ______________
7. ______________
8. ______________
9. ______________
10. ______________

Name ______________________________ Date ______________

Fingerprints

Fingerprints form unique patterns. No two people have the same pattern, not even identical twins. There are three main types of patterns: the arch, the loop, and the whorl.

arch

loop

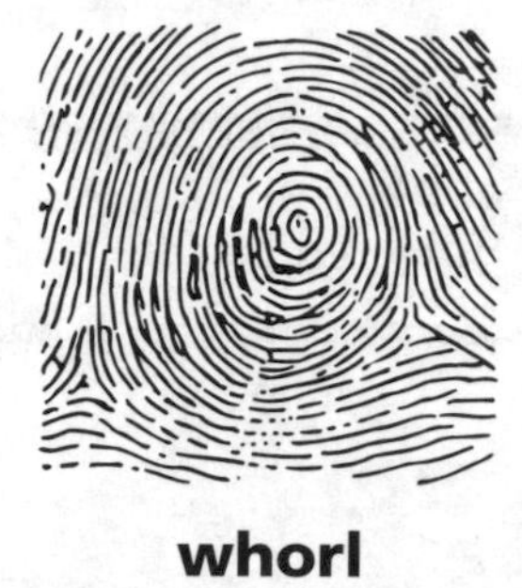

whorl

Make a record of your own fingerprints on the chart below. Then label the type of fingerprint you have on each finger below the boxes.

Step 1: Make sure your hands are clean. Before you begin, cut ten pieces of wide clear tape and stick them carefully to the edge of your desk. Handle the tape carefully so it doesn't get covered with prints.

Step 2: Rub the side of the graphite on a sharpened pencil over the pad of one finger. Then carefully place the sticky side of the tape against the darkened pad of the finger.

Step 3: Carefully peel off the tape and place it on the chart. Continue until you have recorded all ten prints.

Right Hand

________	________	________	________	________

Left Hand

________	________	________	________	________

Name ______________________ Date ____________

Toenails and Fingernails

Nails on the fingers and toes are made of hard, keratin-filled epidermal cells. They protect the ends of the fingers and toes from injury, help us grasp small objects, and enable us to scratch when we are itchy. Use the terms in the word box to label the diagrams. Some words are used more than once.

bone	nail root	nail bed	blood vessel
fatty tissue	nail body	free edge	lunula

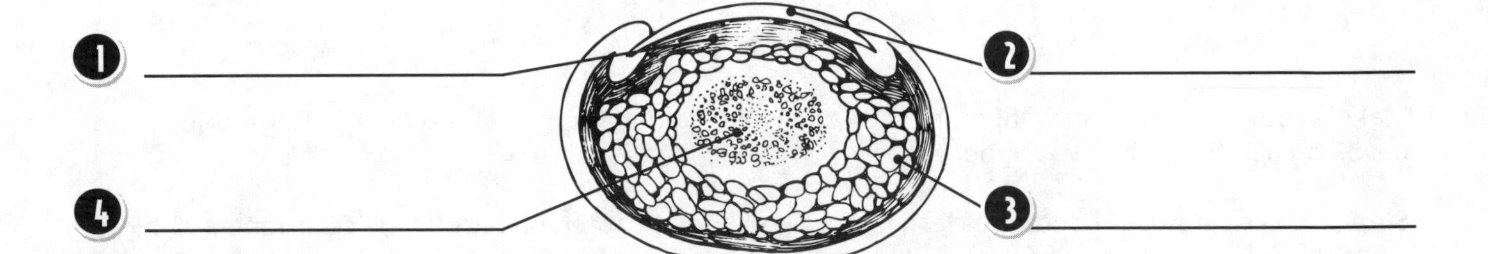

1. ______________________
2. ______________________
3. ______________________
4. ______________________

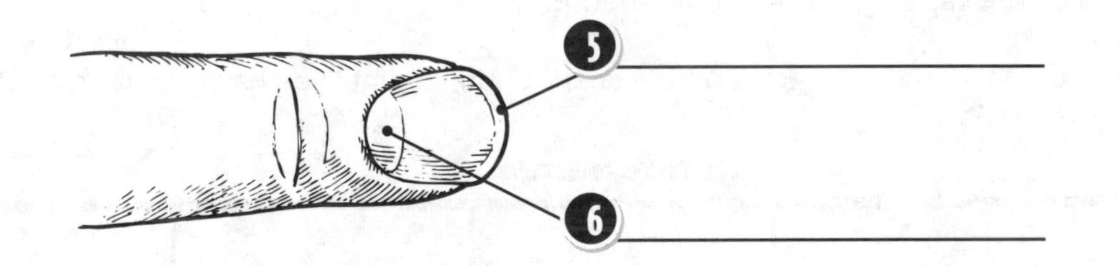

5. ______________________
6. ______________________

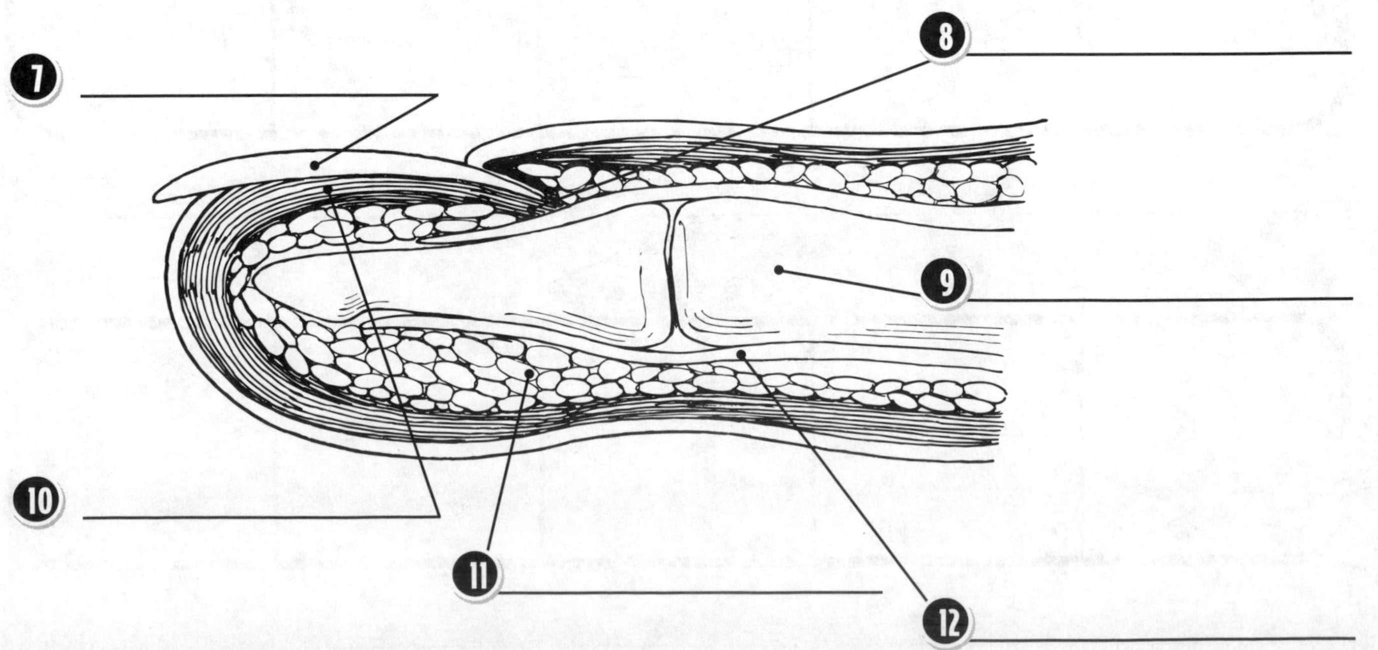

7. ______________________
8. ______________________
9. ______________________
10. ______________________
11. ______________________
12. ______________________

Name ________________________________ Date ______________

Parts of a Nail

As the nail grows, it advances forward by the successive growth of new cells at the root and under the body of the nail. Because the nails themselves are dead tissue, cutting or trimming them does not hurt. Match each term in the word box to its description. Some words are used twice.

bone	nail root	nail bed	blood vessels
fatty tissue	nail body	free edge	lunula

1. ______________ The part of the nail that is visible.
2. ______________ This part of the nail extends past the end of the finger or toe.
3. ______________ Most of the nail body appears pink because of these just underneath.
4. ______________ This semicircular area at the base of the nail is pale because of the thick layer of epidermal cells that don't contain blood vessels.
5. ______________ The fatty tissue and nail help protect this located inside the finger.
6. ______________ This is the growing part of the nail found under the skin.
7. ______________ The part of the visible nail that is attached to the finger.
8. ______________ The layer of fat cells that act as a cushion at the fingertip.

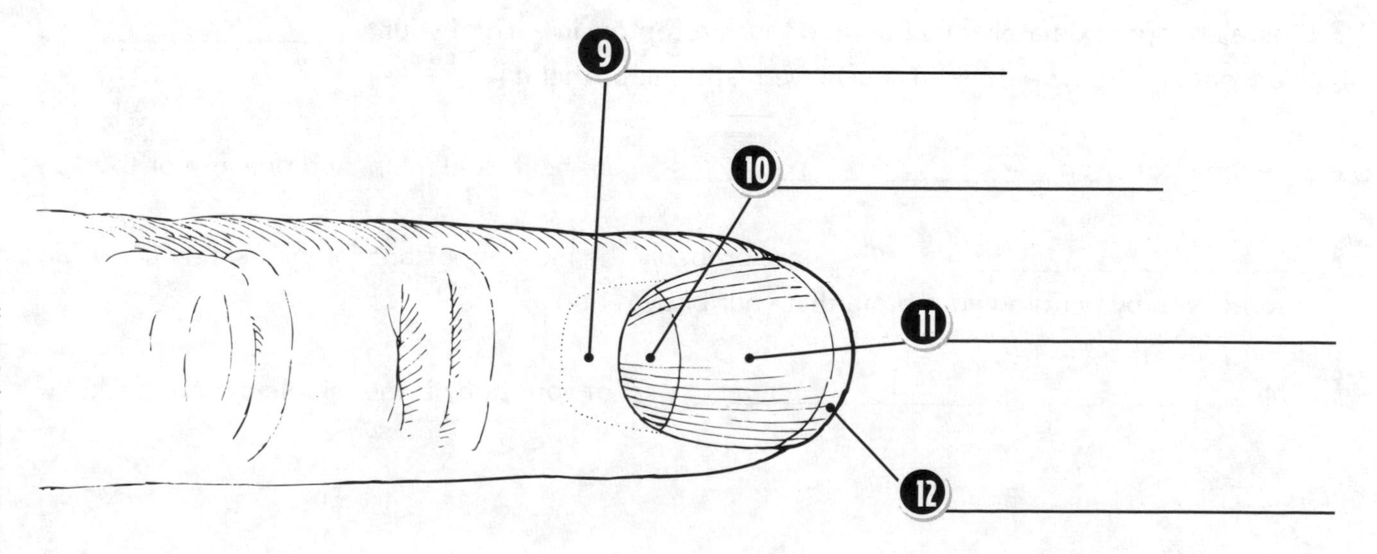

Name ______________________________ Date ______________

The Sense of Touch

Touch is one of the five senses. With it, the body senses it is in contact with other substances. In humans, touch is accomplished by nerve endings in the skin that convey sensations to the brain via nerve fibers. Match each term in the word box to its description.

spinal cord	dermis	pain receptors	brain	tongue
touch receptors	sensitive areas	least sensitive	fingers	

1. The sense of touch originates in the bottom layer of your skin called the ______________.

2. Nerve endings carry information from their location in the skin to the ______________, which in turn relays it to the brain.

3. The ______________ is ultimately the organ that interprets the stimuli as pain, temperature, or pressure.

4. The most ______________ of your body are your hands, lips, face, neck, tongue, fingertips, and feet.

5. Your ______________ has many nerve endings, but they are more sensitive to pain than to hot or cold. This is why it is so easy to burn your mouth when you eat something really hot.

6. There are approximately one hundred touch receptors in each of your ______________. No wonder when we think of touch, we first think of them!

7. The most common ______________ are heat, cold, pain, and pressure or touch.

8. ______________ are probably the most important for your safety. They can protect you by warning your brain that your body is hurt.

9. The ______________ part of your body is the middle of your back.

Name ______________________________ Date ______________

The Reproductive System: Female

The reproductive system refers to those organs that are necessary for the reproductive process. It also includes those organs that distinguish female from male, even though they may not be directly involved in actual reproduction. Use the terms in the word box to label the diagram.

ovary	vagina	uterus	cervix	Fallopian tube

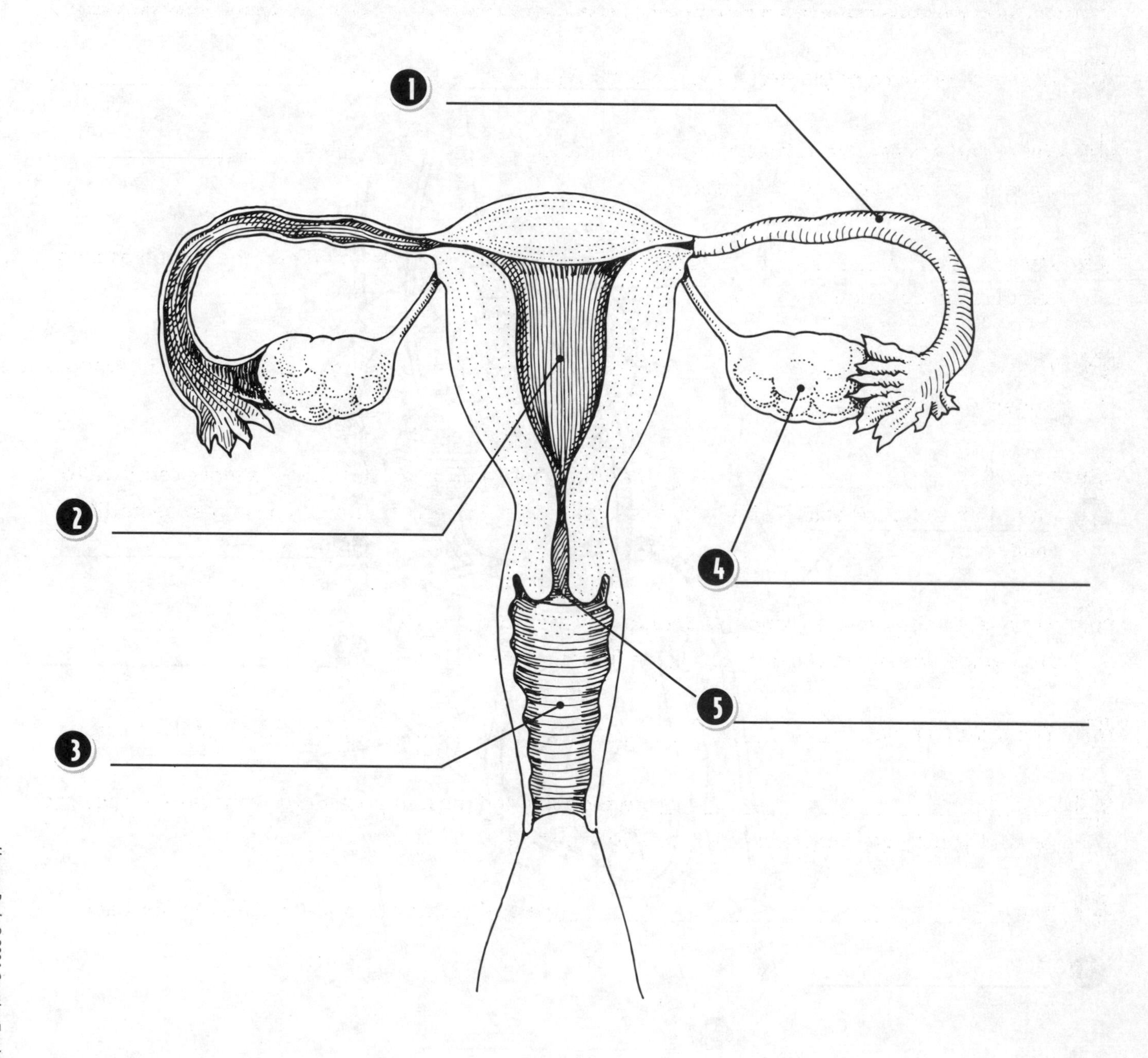

Name ______________________________ Date ______________

The Reproductive System: Male

The reproductive system is the only system in the body where there are distinct differences between male and female. The purpose of the system is to primarily create new life. Use the terms in the word box to label the diagram.

testis	urethra	scrotum	sperm tube	penis	bladder

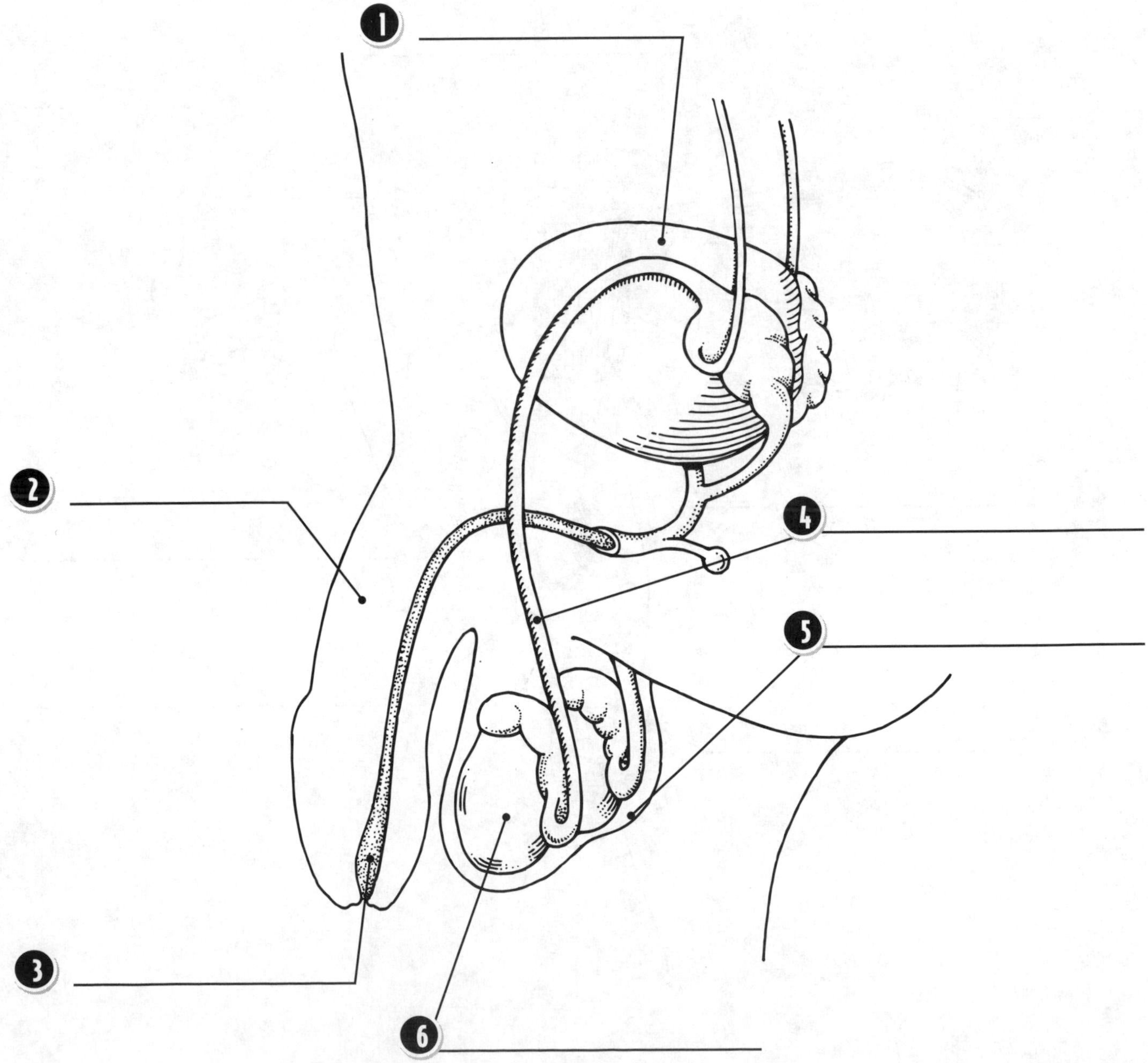

1 ______________

2 ______________

3 ______________

4 ______________

5 ______________

6 ______________

Name ______________________ Date ____________

Role of Reproductive Organs and Hormones

For both males and females there are hormones that regulate the organs and processes of the reproductive system. Each organ has a specialized function in the creation of a new life. Match each term in the word box to its description.

estrogen	progesterone	testosterone	ovary	uterus
vagina	testis	penis	urethra	

1. ______________ Male organ in which sperm and androgen are produced.
2. ______________ This is the birth canal.
3. ______________ The chamber in the female in which the new individual will develop before birth.
4. ______________ The male organ through which sperm passes to the female.
5. ______________ Produces the human egg cell and some female reproductive organs.
6. ______________ A female hormone that prepares the uterus for the implantation of a fertilized egg.
7. ______________ A male hormone produced in the testis. Controls the development and maintenance of masculine characteristics.
8. ______________ Produced by the ovaries, this female hormone helps develop female characteristics.
9. ______________ The canal within the penis through which semen is discharged.

Name ____________________ Date ____________

The Development of a New Life

From the time of conception, a single cell begins the process to become a new living being. This single cell divides again and again until it forms the six trillion cells of a human newborn baby. This entire process takes a period of 38 weeks or nine months. Use the descriptions in the word box to label each stage of development.

fully developed with organs that can function on their own
first movement is felt and heartbeat can be heard with a stethoscope
develops tiny arm and leg buds; heart begins to beat
has recognizable human features; gender can be determined
ears, eyes, nose, fingers, and toes are formed
can survive birth with special care

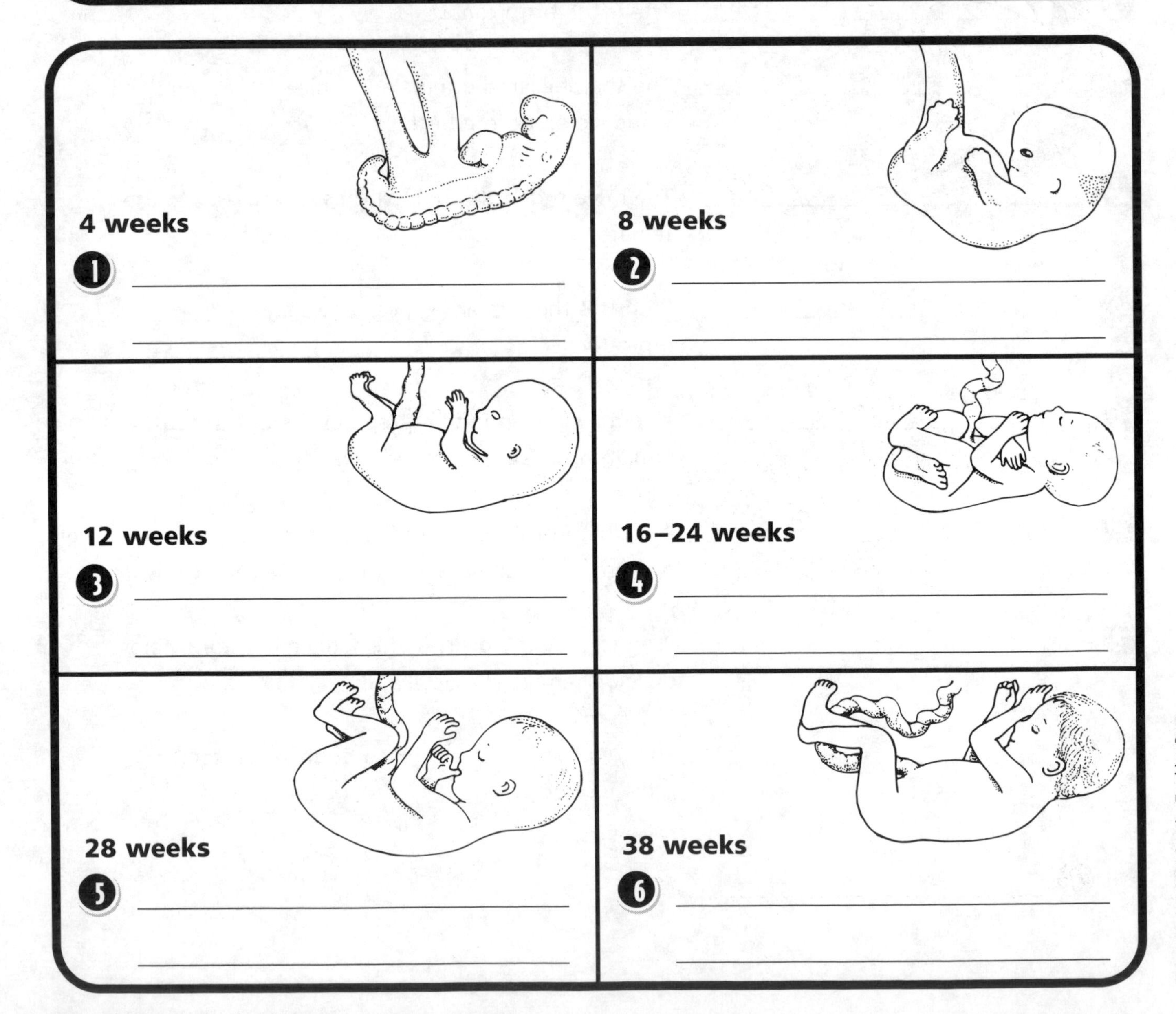

Name ______________________________ Date ______________

Birth of a Baby

Pregnancy normally ends 38 weeks after the time of conception. The birth process is called labor. This process is stimulated by the release of a hormone from the pituitary gland, which starts contractions in the uterus. The cervix eventually enlarges enough to allow the baby to pass through. The amniotic sac that surrounds the baby will break. This results in a gush of amniotic fluid; mothers will say at this point that their "water broke." After the baby is born, the placenta separates from the uterine wall and is pushed out with more contractions. Use the terms in the word box to label the diagrams.

birth canal	placenta	uterus	umbilical cord

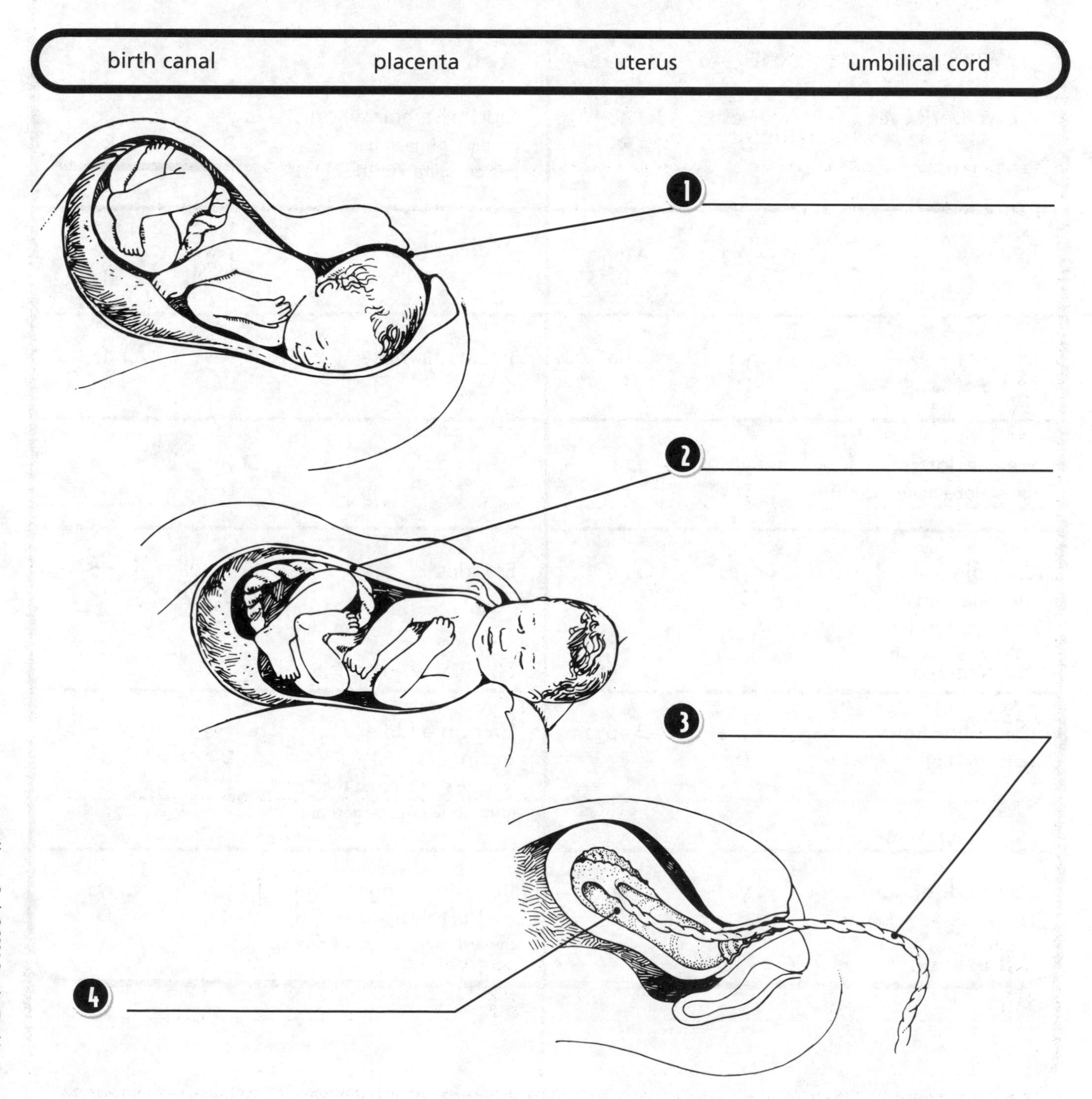

Name ______________________________ Date ______________

Inherited Traits

Genes are units of information about traits, or characteristics, we have inherited from our parents. Many of the traits that make up our physical appearance are inherited from our parents. The chart shows many physical traits that you may or may not have. Check the appropriate box to record your inherited traits.

D= Dominant Trait R= Recessive Trait

Trait	Do you have it?		Trait	Do you have it?	
Brown/Dark eyes	❑ yes D	❑ no R	Clockwise hair whorl (at back of head hair pattern spirals to the right)	❑ yes D	❑ no R
Red hair	❑ yes R	❑ no D	Widow's peak	❑ yes D	❑ no R
Dimples (chin or cheeks)	❑ yes D	❑ no R	Turned-up nose	❑ yes D	❑ no R
Free earlobes (an earlobe hangs down)	❑ yes D	❑ no R	Dark hair	❑ yes D	❑ no R
Ear points (the inner part of ear that curls over has a point of skin on it)	❑ yes D	❑ no R	Freckles	❑ yes D	❑ no R
Can roll tongue (sides roll up)	❑ yes D	❑ no R	Hair on middle of fingers (on the section past the first joint above the fingernail)	❑ yes D	❑ no R
Can fold tongue (tip can be folded up to almost touch nose)	❑ yes R	❑ no D	Bent little fingers (pinkies held up curve inward toward other fingers)	❑ yes R	❑ no D

Name ______________________________ Date ______________

Dominant or Recessive?

Genes come in two strengths, dominant or recessive. Dominant genes are stronger. They overpower recessive genes. If a trait is dominant and present in an individual, that trait will be apparent. Recessive genes are weaker. Recessive traits will only appear if no dominant gene for the same type of trait is present. Categorize the traits in the word box by writing them in the correct column of the chart.

dark-colored eyes
can roll tongue
ear points
can fold tongue
widow's peak
hemophilia
blond hair
dark hair
freckles
hair on middle of fingers
bent little finger
sickle-cell anemia
dimples
red hair
free earlobes
turned-up nose
albinism

Dominant Trait	Recessive Trait
______________	______________
______________	______________
______________	______________
______________	______________
______________	______________
______________	______________
______________	______________
______________	
______________	
______________	

Name ______________________________ Date ____________

Reproductive Review

Use what you have learned about the reproductive system to label each diagram.

ovary	testis	scrotum	penis	vagina	uterus
Fallopian tube	sperm tube	bladder	cervix	urethra	

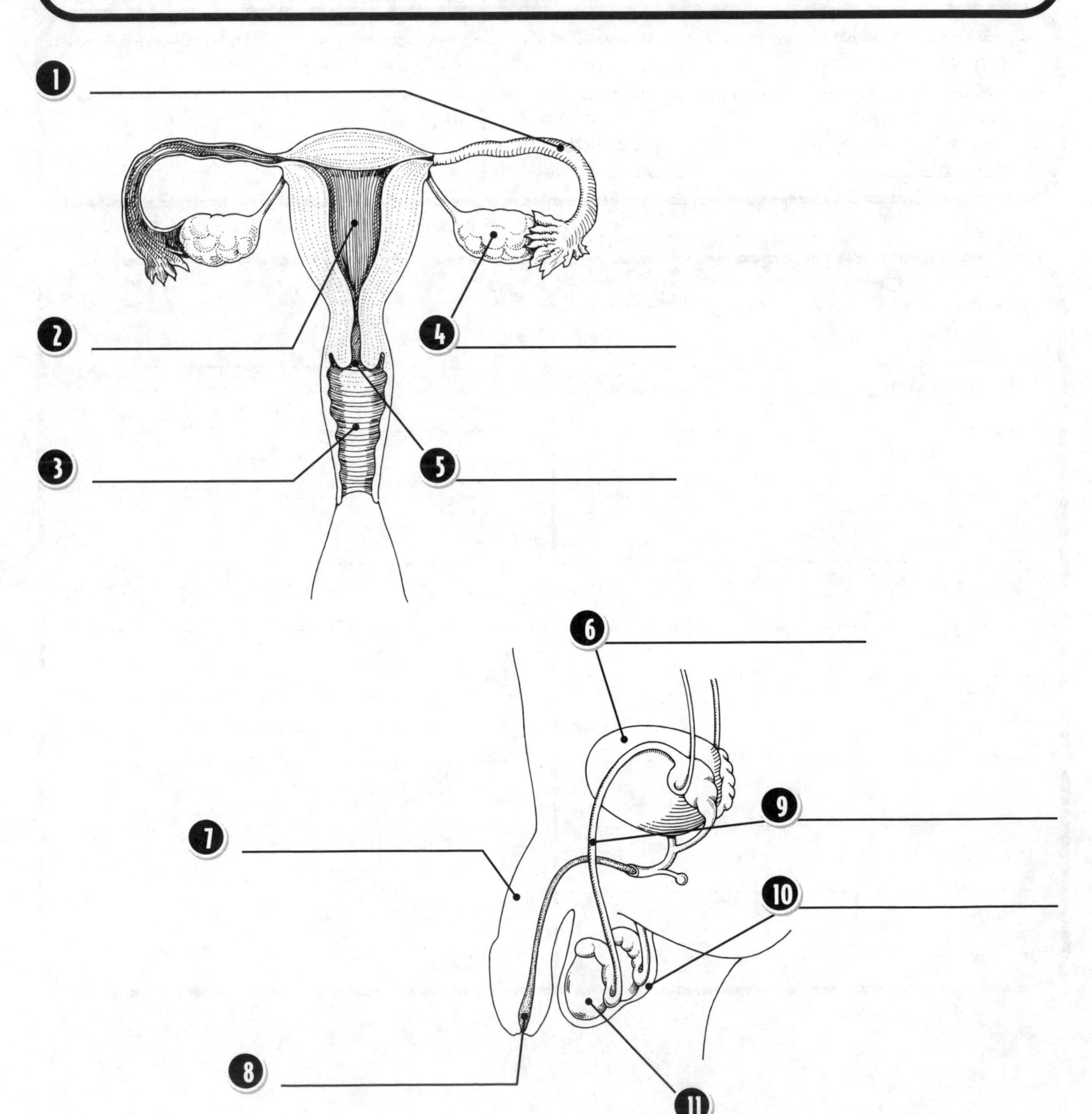

Name ______________________________ Date ____________

Body Nutrients

The cells, tissues, and organs in your body need food to stay alive. Your body requires nutrients from the foods you eat to help the cells grow and repair themselves. Nutrients can be considered nonessential or essential. Nonessential nutrients are those that the body can manufacture itself. Essential nutrients are those that the body must obtain from food. Match each essential nutrient in the word box to its description.

fat	protein	carbohydrate	water	vitamin	mineral

1. ______________________ The body's building material, this nutrient is needed to make new tissue. Milk, beans, meat, and peanuts are good sources.

2. ______________________ This nutrient makes up over half of your body weight. One of its many functions is to carry the other nutrients throughout the body.

3. ______________________ This nutrient helps build strong bones and teeth. It is also necessary for healthy red blood. Sources can be found in all food groups.

4. ______________________ This nutrient provides energy to work and play. Starchy foods like potatoes and pasta have plenty of this nutrient.

5. ______________________ This nutrient is needed to help the body get energy from other nutrients. This nutrient can be divided into smaller groups, each designated a letter of the alphabet.

6. ______________________ This nutrient provides a concentrated source of energy. Healthy sources of this nutrient are nuts, olives, and avocados.

Name ______________________________ Date ______________

Vitamins and Minerals

Vitamins and minerals are organic substances that are needed in small amounts to help the body function properly. Use the vitamins and minerals in the word box to complete the chart.

vitamin A	vitamin D	vitamin E	vitamin C	niacin
calcium	iodine	potassium	iron	magnesium

Vitamin or Mineral	How It Helps the Body	Food Sources
1 ____________	prevents scurvy and enhances the body's resistance to infection	fruits and vegetables, citrus fruits, cabbage, broccoli, green pepper
2 ____________	helps form healthy bones and teeth; promotes blood clotting and nerve transmissions	dairy products, dark green vegetables, dried legumes
3 ____________	prevents goiter	marine fish, shellfish, iodized salt, dairy products
4 ____________	protects fatty acids and cell membranes from oxidation	whole grains, dark green vegetables, vegetable oils
5 ____________	enhances calcium absorption	in sunlight formed in skin, egg yolk, fish liver oils
6 ____________	prevents pellagra and promotes healthy nerves	green leafy vegetables, potatoes, peanuts, poultry, fish, pork, beef
7 ____________	maintains proper fluid balances	bananas, cantaloupe, leafy vegetables
8 ____________	promotes healthy red blood cells	whole grains, leafy green vegetables, nuts, eggs, meat, molasses, dried fruit
9 ____________	allows proper muscle contraction, enzyme action, and healthy nerves	whole grains, legumes, nuts, dairy products
10 ____________	promotes healthy eyesight and skin	yellow fruits, yellow and green leafy vegetables, egg yolk, fish liver

Name ______________________ Date ____________

Preventing Illness

The human body has physical, chemical, and cellular defenses against bacteria, viruses, and other foreign invaders. But to help the body heal itself, humans have also discovered ways to prevent and cure illnesses. Match each term in the word box to its description.

vaccine	antigen	antibody	bacteria	virus
smallpox	immunization	allergy	polio	white blood cells

1. ______________________ A highly infectious disease that mainly affects children. At its worse, it can cause paralysis and deformity. Through vaccination, it is preventable.

2. ______________________ A simple microscopic agent, consisting of DNA or RNA and a protein coat, that can cause infection or illness.

3. ______________________ Processes that promote the body's immunity against specific disease.

4. ______________________ A very infectious, often fatal disease caused by a virus. Those who survive are often disfigured with scars from the sores. A vaccine is used to prevent this disease.

5. ______________________ A single-cell microorganism. Some are necessary for body processes. Others can cause illness and disease.

6. ______________________ Part of the circulatory system, these help protect the body from infection and disease.

7. ______________________ An unusually high sensitivity to certain substances that trigger a reaction, such as sneezing or rash.

8. ______________________ A substance that when introduced into the body triggers the production of an antibody.

9. ______________________ A swallowed or injected preparation that increases immunity to certain diseases.

10. ______________________ A substance secreted into the lymph or blood in response to the presence of a bacteria, virus, parasite, or other foreign material.

Name ______________________ Date ____________

Pressure Points

When a person is severely cut and begins to bleed, help is quickly needed. First aid for severe bleeding involves applying pressure over the wound. Sometimes using a pressure point can help stop the bleeding, too. A pressure point is a spot above the wound where the artery can be pressed against bone to slow blood flow. A person's pulse can also be felt clearly at pressure points. Draw an arrow to and label each of the pressure points listed in the word box.

neck	wrist	behind the knee	bend of elbow	inside of thigh	top of foot

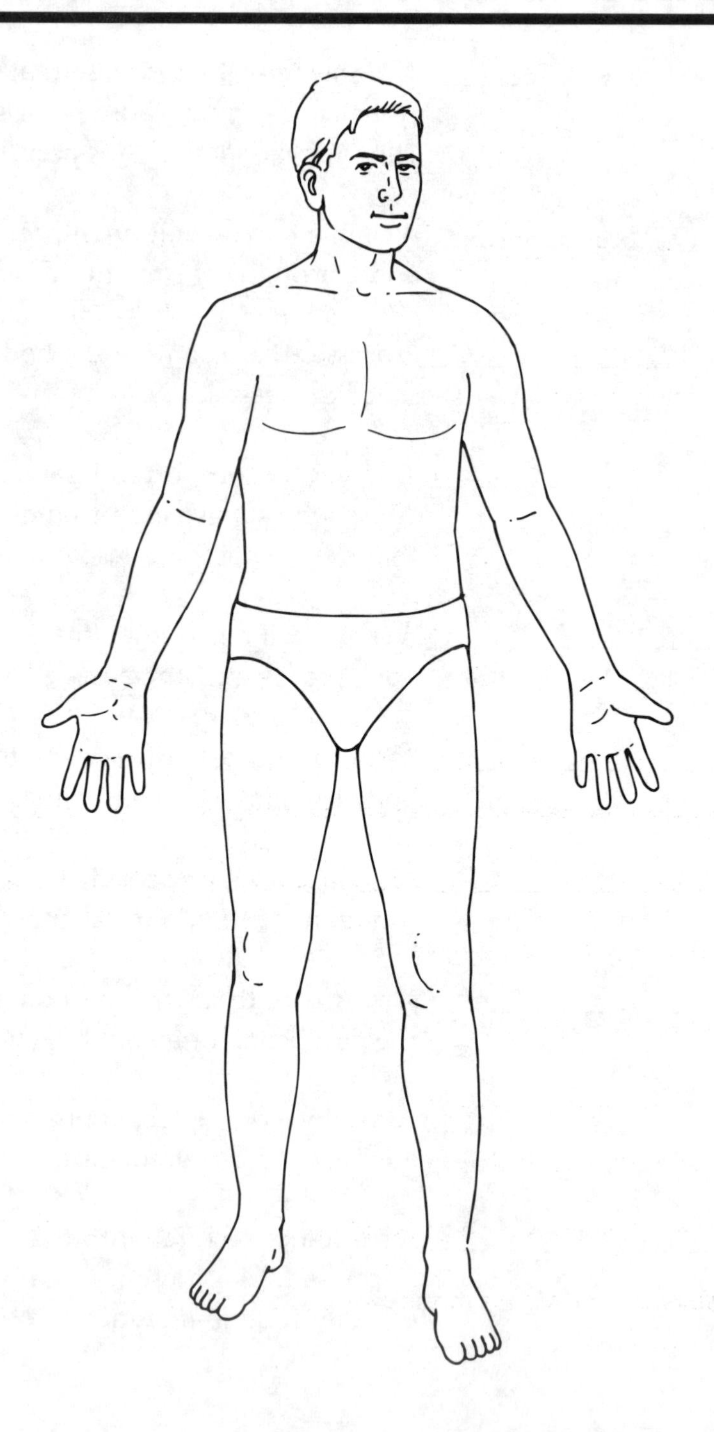

Name ______________________________ Date ______________

The Heimlich Maneuver

The Heimlich maneuver is used to dislodge food that has become stuck in a person's trachea. Use the terms in the word box to label the diagram.

navel	base of rib cage	trachea	food lodged in trachea	fist

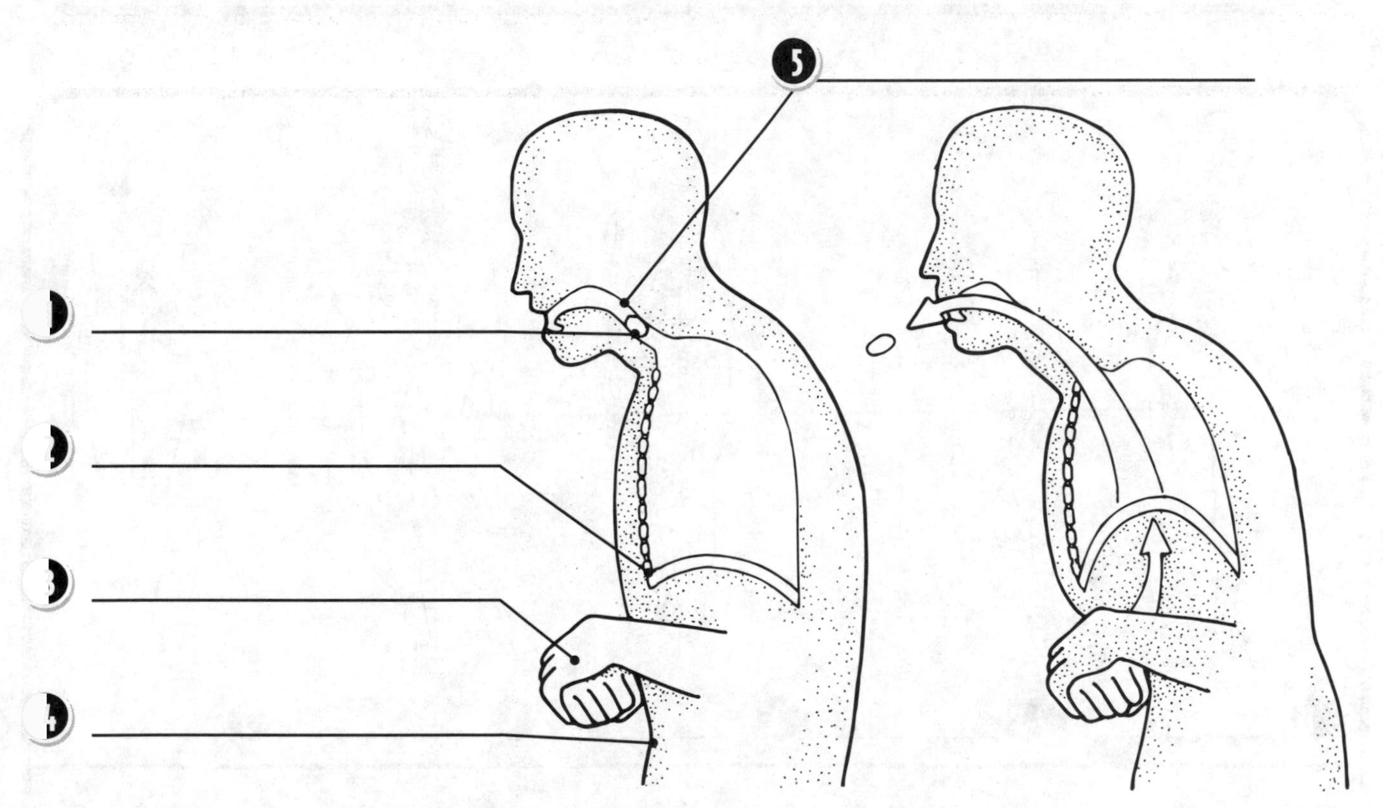

Use the words in the word box to complete the steps of the Heimlich maneuver.

lying down	fist	thumb	trachea	navel	thrust

6. When food becomes lodged in a person's ______________________, it can impair breathing.
7. Stand behind and place arms around the victim. Make a ____________________ with one hand.
8. Position the ____________________ of your fist against the victim's abdomen.
9. Your fist should be slightly above the ____________________ but well below the rib cage.
10. Press your fist into the abdomen with a sudden upward ______________________.
11. Repeat the thrust several times. The maneuver can be performed on someone who is standing, sitting, or ____________________________.

Name ______________________________ Date ______________

All Body Systems

Use the terms in the word box to label the illustration of each body system.

skeletal	respiratory	nervous	muscular	circulatory
lymphatic	digestive	urinary	endocrine	

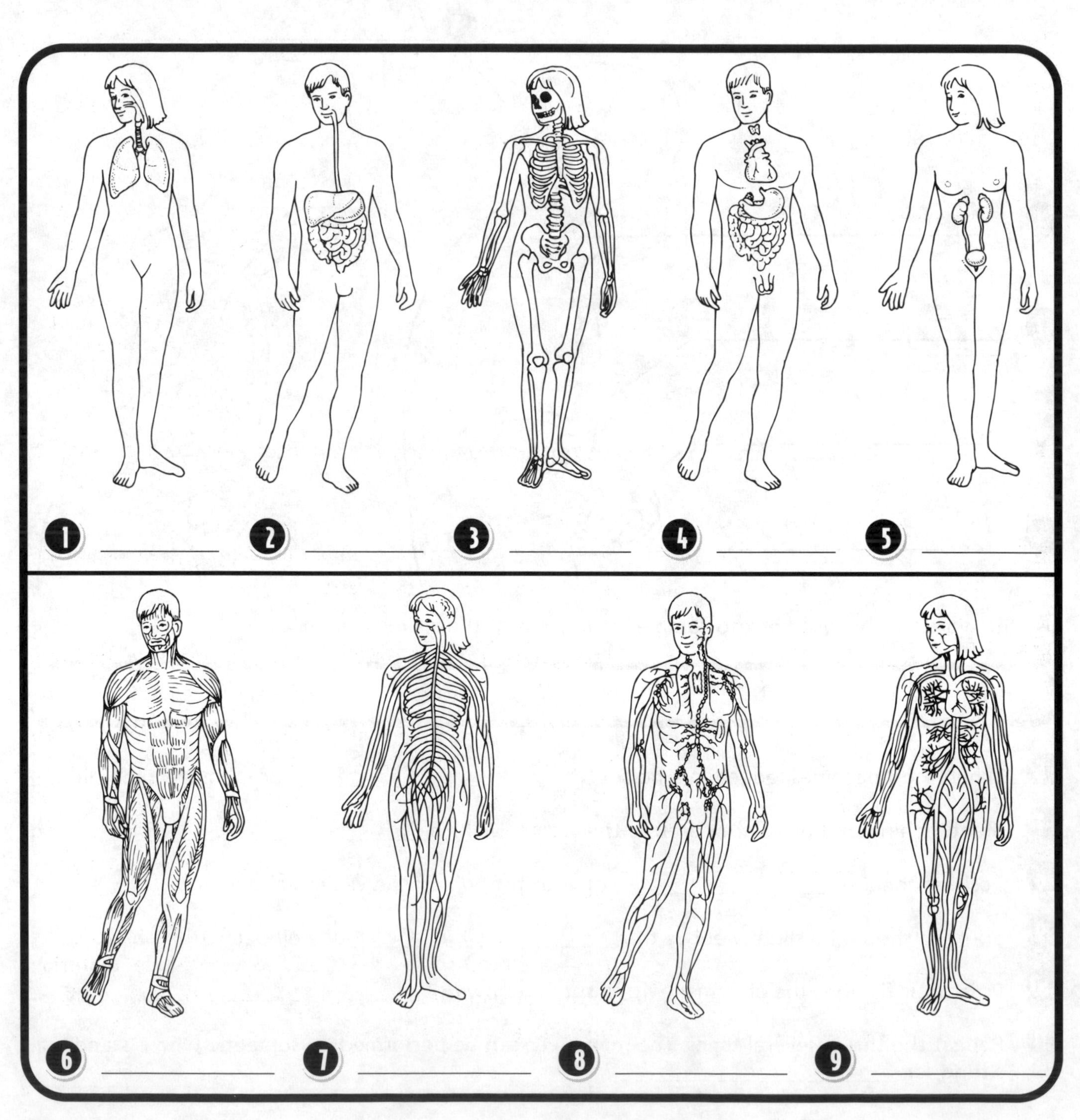

Name ______________________________ Date ______________

Identify the Systems

Match each term in the word box to its description.

skeletal	respiratory	nervous	muscular	circulatory
reproductive	digestive	urinary	endocrine	lymphatic

1. ______________________ This system acts as a means of support for the entire body.
2. ______________________ This system works with the circulatory system to transport fluids and filter out bacteria and other foreign invaders.
3. ______________________ This system breaks down food into the nutrients it needs to function. It also carries indigestible food out of the body as waste.
4. ______________________ The main purpose of this system is to create a new living individual.
5. ______________________ This system contains organs that produce, collect, and eliminate urine from the body.
6. ______________________ This system controls movement of the body, the organs, and the beating of the heart.
7. ______________________ This system relays electrical impulses to and from the brain.
8. ______________________ The organs in this system are involved in the exchange of oxygen and carbon dioxide between the body and its environment.
9. ______________________ This is a system of glands that help body functions by the release of hormones.
10. ______________________ This system pumps nutrients and other essential materials to the cells and removes waste products from the cells through a system of vessels.

Name ______________________ Date ______________

Classify the Parts

Classify each organ or body part in the word box by writing it under the correct heading.

malleus	aorta	auricle	duodenum	red blood cell
ventricle	alveoli	trachea	spleen	thymus gland
retina	pupil	deltoid	diaphragm	biceps
tendon	trapezius	pleura	lymph node	colon
tonsils	stomach	capillary	esophagus	

Circulatory	Lymphatic
____________	____________
____________	____________
____________	____________
____________	____________
Digestive	**Muscular**
____________	____________
____________	____________
____________	____________
____________	____________
Respiratory	**Sensory**
____________	____________
____________	____________
____________	____________
____________	____________

Name ______________________________ Date ______________

Body Parts

We use common terms to refer to various parts of our body's surface. Use the terms in the word box to label the diagram.

forehead	nose	chest	abdomen	calf	forearm
thumb	hip	instep	shin	sole	cheek
palm	thigh	shoulder	heel		

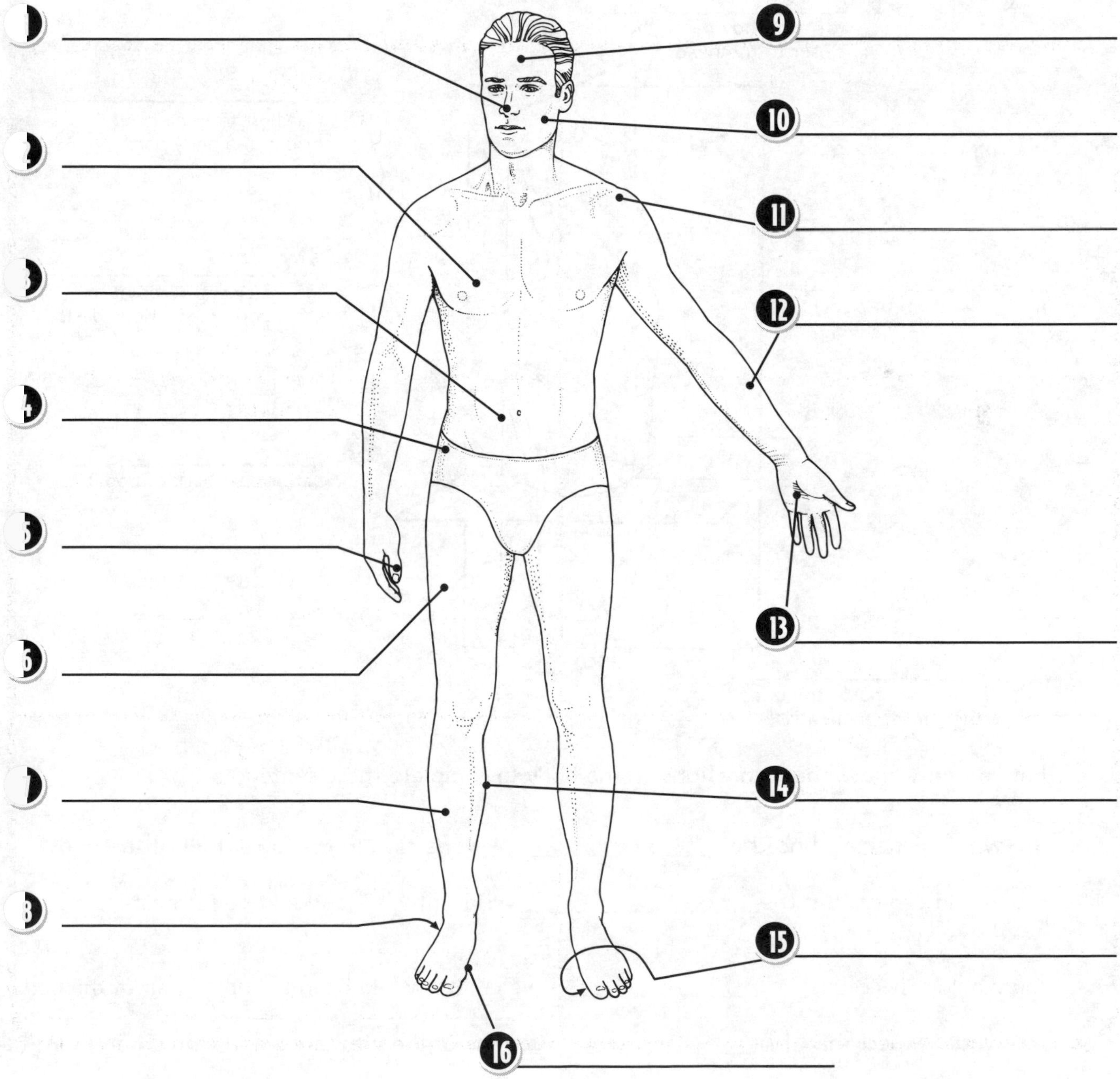

1. ______________
2. ______________
3. ______________
4. ______________
5. ______________
6. ______________
7. ______________
8. ______________
9. ______________
10. ______________
11. ______________
12. ______________
13. ______________
14. ______________
15. ______________
16. ______________

Name ______________________________ Date ______________

Scientific Sections

Words that describe the body, in scientific terms, often get part of their name from where they are found in the body. Imagine that the body is divided front to back and side to side, as shown in the diagram. The use of the terms in the word box will give clues to where the part is found in the body. Use the terms in the word box to label the diagram.

anterior	inferior	posterior	superior	distal	proximal

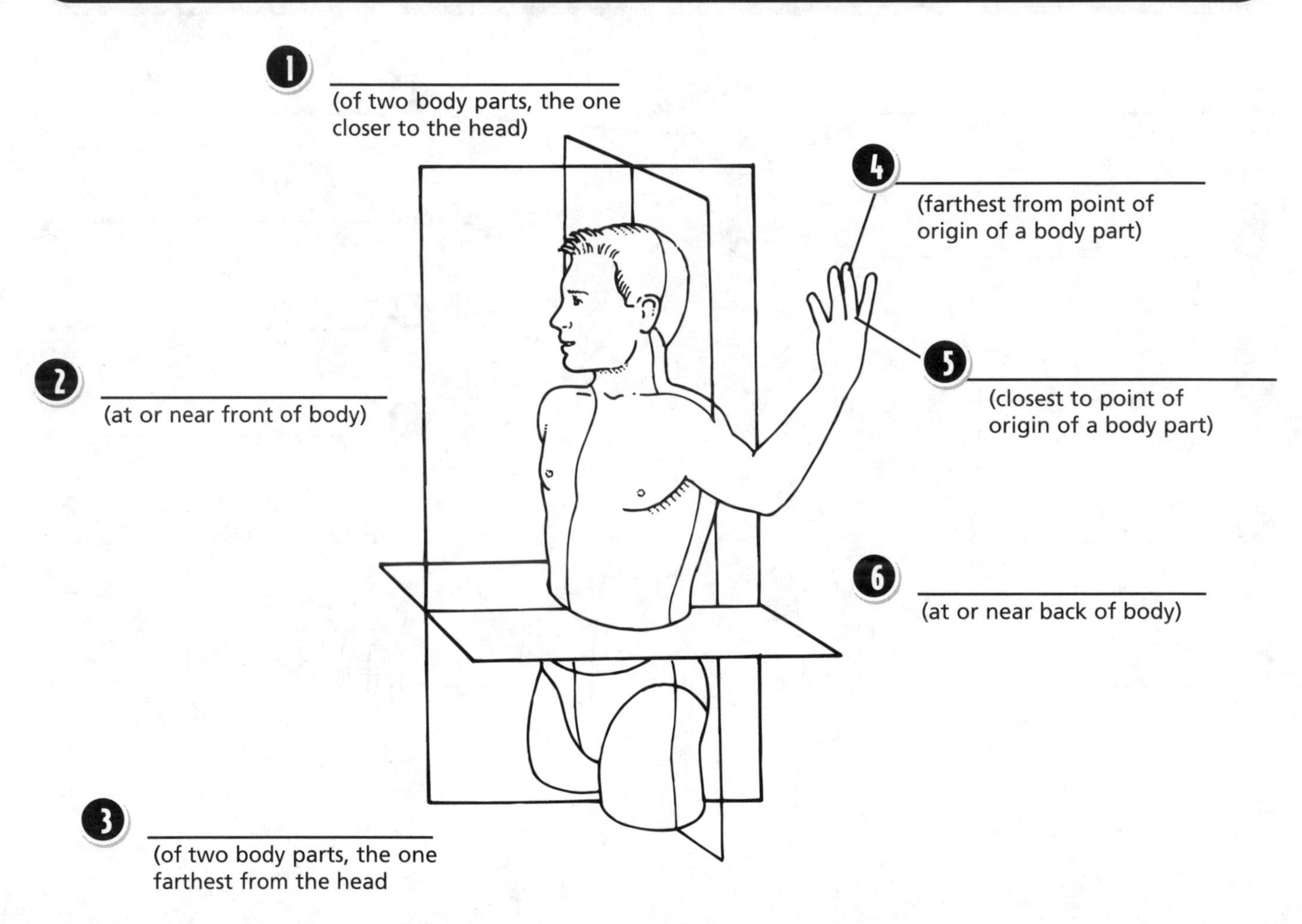

Use what you now know about locations on the body to complete these sentences.

7. You would expect to find the ____________ vena cava in the lower half of the body.
8. You would expect that the ____________ end of the tibia would be farthest from the shoulder.
9. You would expect that the ____________ vena cava would be in the upper half of the body.
10. You would expect that the ____________ lobe of the pituitary gland would be the front.

Name ____________________ Date __________

Functions of Organs

Each organ in the body fits a specific purpose as part of a system. Match each term in the word box to its description.

heart	skull	pancreas	spleen	lung
medulla	esophagus	kidney	bladder	colon

1. ____________________ Part of the digestive system, it collects solid waste and removes extra water.
2. ____________________ Part of the circulatory system, it is a muscular pump that moves fluid through a system of tubes.
3. ____________________ Part of the skeletal system, it provides protection for the brain and sensory organs.
4. ____________________ Part of the digestive and endocrine systems, it regulates glucose levels in the blood and secretes fluids for digestion.
5. ____________________ Part of the respiratory system, it removes carbon dioxide from the blood and adds oxygen in its place.
6. ____________________ Part of the lymphatic system, it disintegrates old blood cells and filters foreign substances from the blood.
7. ____________________ Part of the urinary system, it stores urine.
8. ____________________ Part of the digestive system, it contracts to push food into the stomach.
9. ____________________ Part of the central nervous system, it controls respiration, circulation, and other body functions.
10. ____________________ Part of the urinary system, it maintains proper water balance and filters the blood.

Name ______________________ Date ____________

Grouping by Function

Each human body system can be classified by its overall function and purpose in the body. Use the terms in the word box to complete the chart.

skeletal system	respiratory system	nervous system	muscular system
circulatory system	digestive system	urinary system	endocrine system
lymphatic system	sensory system		

Control Group

____

____

____

Movement Group

____

____

Energy Group

____

____

____

____

____

Name ______________________ Date ______________

Human Body Review

Review what you have learned about the human body. Match each term in the word box to its definition.

adrenal	biceps	bladder	cervix	epidermis
excretory	enamel	fracture	genes	heart
ovaries	palm	pancreas	pelvis	pituitary
pressure point	pulse	testes	sebum	tendons

1. ______________ waste removal system
2. ______________ male glands that produce hormones
3. ______________ outer layer of the tooth
4. ______________ outer layer of skin
5. ______________ these attach muscles to the skeleton
6. ______________ a break in a bone
7. ______________ location on body to feel pulse
8. ______________ female glands that produce hormones
9. ______________ carry information about an individual's traits
10. ______________ metatarsals of the hand
11. ______________ opening to the uterus
12. ______________ upper arm muscle
13. ______________ gland that controls the body's use of glucose
14. ______________ heartbeat
15. ______________ framework of bones that supports lower abdomen
16. ______________ oily substance given off by the sebaceous gland
17. ______________ blood pump
18. ______________ stores urine
19. ______________ controls body growth and function of other glands
20. ______________ gland that goes to work when we are frightened or excited

Answer Key

The Human Cell (page 5)

1. mitochondrion
2. endoplasmic reticulum
3. nucleus
4. vacuole
5. Golgi body
6. cell membrane
7. cytoplasm

Types of Cells (page 6)

1. red blood cells
2. nerve cells
3. cartilage cells
4. skeletal muscle cells
5. adipose cells
6. bone cells

Types of Tissue (page 7)

1. nerve tissue
2. nerve fiber
3. connective tissue
4. collagen
5. epithelial tissue
6. nucleus
7. muscle tissue
8. cell

The Skeletal System (page 8)

1. scapula
2. rib cage
3. vertebrae
4. pelvis
5. patella
6. cranium
7. mandible
8. clavicle
9. radius
10. coccyx
11. femur
12. tibia

Common Names for Fancy Words (page 9)

(see page 126)

Bones of the Hands and Feet (page 10)

(see page 126)

Bones of the Leg and Arm (page 11)

1. femur
2. patella
3. fibula
4. tibia
5. humerus
6. ulna
7. radius

Making Connections (page 12)

Ball-and-socket	Gliding	Hinge	Pivot
shoulder	wrist	elbow	forearm
hip	ankle	knee	neck
	backbone	toe	
		finger	

This Bone Is Connected to That Bone (page 13)

1. neck
2. toe
3. wrist
4. thighbone
5. backbone
6. shinbone
7. shoulder blade
8. palm
9. head
10. instep

Scientifically Speaking (page 14)

1. cervical
2. phalanges
3. tarsals
4. femur
5. patella
6. sternum
7. vertebrae
8. ulna
9. humerus
10. tibia

Parts and Function of a Bone (page 15)

1. periosteum
2. blood vessels
3. marrow
4. spongy bone
5. compact bone
6. cartilage
7. marrow
8. periosteum
9. spongy bone
10. blood vessels
11. cartilage
12. compact bone

Your Teeth (page 16)

1. crown
2. neck
3. root
4. enamel
5. dentin
6. pulp
7. cementum
8. root canal
9. nerve
10. incisors
11. canines
12. bicuspids
13. molars

Bones of the Head (page 17)

1. parietal bone
2. frontal bone
3. maxillary nasal bone
4. nose cartilage
5. teeth
6. mandible
7. vertebrae
8. occipital bone
9. cranium
10. temporal bone
11. eye socket

Injuries to Bones (page 18)

1. multiple
2. spiral
3. greenstick
4. closed
5. open
6. comminuted

The Backbone (page 19)

1. cervical region
2. thoracic region
3. lumbar region
4. disc
5. vertebra
6. sacrum
7. coccygeal region
8. neck
9. tailbone
10. chest
11. lower back
12. pelvic girdle

The Human Pelvis (page 20)

1. sacroiliac joint
2. sacrum
3. hip bone
4. coccyx
5. interpubic joint

What Do You Know About Bones? (page 21)

1. metatarsals
2. patella
3. dentin
4. thoracic region
5. periosteum
6. ball-and-socket
7. mandible
8. femur
9. bicuspid
10. scapula
11. clavicle
12. ulna

The Muscular System (page 22)

1. sternocleidomastoids
2. deltoids
3. pectorals
4. intercostals
5. biceps
6. triceps
7. quadriceps
8. gastrocnemius

Work Those Muscles! (page 23)

1. biceps
2. thigh muscles
3. shoulder muscles
4. gluteus maximus
5. calf muscles
6. abdominals
7. neck muscles
8. triceps
9. trapezius
10. chest muscles

Types of Muscle Tissue (page 24)

1. smooth muscles — move food through digestive tract; squeeze the bladder; contract blood vessels

2. cardiac muscles — maintain a heartbeat; keep blood pumping; found only in the heart

3. skeletal muscles — bend arms and legs; close a fist; create a smile or a frown

Muscular Connections (page 25)

1. biceps
2. triceps
3. biceps relaxed
4. triceps contracted
5. tendon
6. biceps contracted
7. triceps relaxed
8. hamstring relaxed
9. quadriceps contracted
10. quadriceps relaxed
11. hamstring contracted

Muscular Actions (page 26)

1. muscle cramp
2. muscle tone
3. muscle spasm
4. atrophy
5. muscular dystrophy
6. reflex
7. hypertrophy
8. muscle ache

The Circulatory System (page 27)

1. lungs
2. heart
3. liver
4. kidneys
5. artery
6. vein
7. capillaries

Veins and Arteries (page 28)

(see page 127)

The Heart (page 29)

1. pulmonary veins
2. vena cava
3. aorta
4. right atrium
5. right ventricle
6. pulmonary artery
7. left atrium
8. left ventricle

The Heart Has a Job to Do (page 30)

1. right lung
2. right atrium
3. right ventricle
4. lower body
5. left ventricle
6. left lung
7. left atrium
8. upper body

What's in Your Blood? (page 31)

1. white blood cells
2. leukocyte
3. platelets
4. blood volume
5. plasma
6. red blood cells
7. 4–5 quarts
8. hemoglobin
9. water
10. fibrinogen

Go with the Flow, Part I (page 32)

1. right atrium
2. right valve
3. ventricle
4. valve
5. pulmonary
6. vein
7. left atrium
8. left ventricle
9. aorta
10. inferior vena cava

Go with the Flow, Part II (page 33)

(see page 127)

Circulatory Review (page 34)

1. aorta
2. right atrium
3. right ventricle
4. pulmonary artery
5. left atrium
6. left ventricle
7. lungs
8. heart
9. liver
10. kidneys

The Respiratory System (page 35)

1. pharynx
2. bronchial tube
3. pleura
4. nasal cavity
5. mouth
6. larynx
7. trachea
8. diaphragm
9. voice box
10. lung cover
11. windpipe
12. throat

The Lungs (page 36)

1. bronchial tube
2. trachea
3. left lung
4. right lung
5. alveoli
6. pleura
7. lobe
8. bronchiole
9. capillaries
10. alveolar sac

Breathe In, Breathe Out (page 37)

Inhalation

diaphragm is contracted
oxygen goes in
rib cage expands
diaphragm flattens

Exhalation

diaphragm is relaxed
carbon dioxide goes out
rib cage returns to resting position
diaphragm moves up

When Lungs Break Down (page 38)

1. bronchitis
2. lung cancer
3. pneumonia
4. pulmonary embolism
5. pneumothorax
6. emphysema
7. pulmonary edema
8. asthma
9. smoker's cough

Respiratory Review (page 39)

Lung Disorders

bronchitis
emphysema
asthma
pneumothorax

Parts of a Lung

alveoli
pleura
lobe
bronchiole

Respiratory System

bronchial tube
trachea
diaphragm
lung

The Digestive System (page 40)

1. teeth
2. tongue
3. liver
4. gallbladder
5. colon
6. small intestine
7. anus
8. mouth
9. salivary glands
10. esophagus
11. stomach
12. pancreas

The Alimentary Canal (page 41)

1. food enters
   organ: mouth
2. squeezes food down to stomach
   organ: esophagus
3. holds food while digesting
   organ: stomach
4. bile enters
5. pancreatic enzymes enter
6. nutrients absorbed into bloodstream
   organ: small intestine
7. water passes into the bloodstream
   organ: large intestine
8. solid waste exits
   organ: anus

The Stomach (page 42)

1. esophagus
2. circular muscle
3. sphincter
4. duodenum
5. serosa
6. longitudinal muscle
7. oblique muscle
8. mucous membrane

Digestion Helpers (page 43)

1. liver
2. gallbladder
3. bile duct
4. pancreas
5. duodenum
6. bile
7. gallbladder
8. pancreatic juice
9. Brunner's glands
10. enzymes

Mouthing Off! (page 44)

1. palate
2. teeth
3. tongue
4. lip
5. salivary glands
6. uvula
7. pharynx
8. epiglottis
9. esophagus

The Intestines (page 45)

The Small Intestine

three parts: duodenum, jejunum, and ileum
most nutrients are digested and absorbed
about 11 feet long
receives secretions from liver, pancreas, and gallbladder
delivers unabsorbed material to the colon

The Colon

three parts: ascending, transverse, and descending
stores undigested food
about 5 feet long
absorbs minerals and water
also known as the large intestine
where the appendix is attached

Digesting What You Have Learned (page 46)

1. tongue
2. liver
3. esophagus
4. colon
5. small intestine
6. duodenum
7. salivary glands
8. anus
9. stomach
10. pancreas

The Urinary System (page 47)

1. vein
2. kidney
3. bladder
4. artery
5. ureter
6. muscle
7. urethra

Two of a Kind (page 48)

1. renal capsule
2. kidney medulla
3. kidney cortex
4. renal artery
5. renal vein
6. ureter

Maintaining a Balance of Fluids (page 49)

1. thirst behavior
2. blood
3. urine
4. interstitial fluid
5. kidney stones
6. extracellular fluid
7. dialysis
8. hypertension

The Lymphatic System (page 50)

1. tonsils
2. lymphatic duct
3. thymus gland
4. thoracic duct
5. spleen
6. lymph vessels
7. lymph nodes
8. bone marrow

Functions of the Lymphatic System (page 51)

1. tonsils
2. thymus gland
3. spleen
4. bone marrow
5. lymph nodes
6. lymph vessels
7. thoracic duct
8. lymphatic ducts

Working as a Team (page 52)

Lymphatic System

spleen
nodes
lymphocytes
lymph
produces antibodies
transports fats and pathogens
releases lymphocytes

Circulatory System

heart
capillaries
white blood cells
blood
produces red blood cells
transports oxygen and nutrients
forms platelets for clotting

Urinary System

kidney
ureters
urine
filters water and solutes
maintains fluid balance
triggers thirst behavior

The Endocrine System (page 53)

1. pituitary gland
2. thymus
3. adrenal gland
4. ovaries (female)
5. pineal gland
6. thyroid gland
7. pancreas
8. testes (male)

A Particular Gland for a Particular Function (page 54)

1. pituitary
2. pancreas
3. testes
4. adrenal
5. thyroid
6. ovaries
7. parathyroid
8. pineal gland
9. thymus

Producing Hormones (page 55)

1. ovary	regulates female functions
2. pineal gland	regulates sleep-wake cycle
3. pancreas	regulates glucose levels
4. adrenal gland	response to stress
5. thyroid	regulates growth
6. testis	regulates male functions
7. parathyroid	regulates calcium in blood
8. pituitary	regulates pain response

Where Do Those Hormones Come From? (page 56)

1. endorphins
2. parathyroid hormone
3. adrenaline
4. estrogen
5. melatonin
6. thyroxine
7. insulin
8. testosterone

The Central Nervous System (page 57)

1. brain
2. cerebrum
3. cerebellum
4. brain stem
5. spinal cord
6. nerves
7. nerve cell

Neurons (page 58)

1. dendrites
2. cell body
3. terminal fibers
4. nucleus
5. axon

Impulse Transmitters (page 59)

1. transmitting molecule
2. synapse
3. axon
4. axon terminal
5. synaptic cleft
6. dendrite

The Brain (page 60)

1. left cerebral hemisphere
2. right cerebral hemisphere
3. pineal gland
4. thalamus
5. hypothalamus
6. cerebrum
7. cerebellum
8. medulla
9. spinal cord

The Lobes of the Brain (page 61)

1. motor cortex
2. frontal lobe
3. temporal lobe
4. somatosensory cortex
5. parietal lobe
6. occipital lobe
7. frontal lobe
8. temporal lobe
9. somatosensory cortex
10. motor cortex
11. parietal lobe
12. occipital lobe

Autonomic versus Somatic Nervous Systems (page 62)

Autonomic Nervous System

heart beats
pancreas secretes
produces bile
stomach digests
colon contracts
get goosebumps
releases adrenaline
relaxes diaphragm

Somatic Nervous System

fist clenches
fingers curl
climb stairs
make a peace sign
curl tongue
stretch arms
feel a hot pan
point toes

Autonomic Nervous System (page 63)

1. trachea
2. heart
3. liver
4. gallbladder
5. colon
6. rectum
7. eye
8. lungs
9. stomach
10. pancreas
11. small intestine

Nervous System Functions (page 64)

(see page 128)

Central Nervous System Review (page 65)

1. brain stem
2. spinal cord
3. cerebrum
4. cerebellum
5. medulla
6. axon
7. nucleus
8. neuron
9. dendrite

Sensory Systems (page 66)

1. sound processing
2. taste processing
3. sight processing
4. smell processing
5. ear
6. skin
7. eye
8. nose
9. tongue
10. receptor nerve cell
11. sensory nerve cell

The Ear (page 67)

1. auricle
2. wax gland
3. semicircular canals
4. auditory nerve
5. eardrum
6. hammer
7. auditory canal
8. anvil
9. stirrup
10. oval window
11. Eustachian tube
12. cochlea

Functions of the Ear Parts (page 68)

1. stirrup
2. semicircular canals
3. auditory canal
4. hammer
5. oval window
6. eardrum
7. cochlea
8. auricle
9. Eustachian tube
10. anvil

The Outer Ear (page 69)

1. auricle
2. hairs
3. wax gland
4. cerumen
5. auditory canal
6. temporal bone
7. tympanic membrane
8. inner ear
9. middle ear
10. outer ear

The Middle Ear (page 70)

1. malleus
2. Eustachian tube
3. incus
4. stapes
5. inner ear
6. middle ear
7. outer ear

The Inner Ear (page 71)

1. semicircular canals
2. oval window
3. cochlea
4. auditory nerve
5. inner ear
6. middle ear
7. outer ear

What's the Connection? (page 72)

1. nasal passage
2. palate
3. inner ear
4. pinna
5. Eustachian tube
6. epiglottis
7. esophagus
8. trachea

The Eye (page 73)

1. cornea
2. iris
3. aqueous humor
4. lens
5. vitreous body
6. retina
7. optic nerve
8. optic disk
9. sclera
10. iris
11. pupil

Parts of the Eye (page 74)

1. vitreous body
2. lens
3. optic nerve
4. iris
5. pupil
6. sclera
7. optic disk
8. aqueous humor
9. retina

Inside the Eye (page 75)

1. aqueous humor
2. iris
3. pupil
4. cornea
5. lens
6. ciliary muscles
7. vitreous body
8. sclera
9. choroid
10. retina
11. optic nerve
12. optic disk
13. rod cells
14. cone cells

I See, I Understand (page 76)

1. lens
2. cornea
3. image of object
4. optic nerve
5. upside-down image
6. retina
7. visual cortex

The Protection Connection (page 77)

1. eyelids 2. eyelashes
3. conjunctiva 4. tear duct
5. blinking 6. eyebrows
7. skull 8. frontal bone
9. cheekbone

Protecting the Eye (page 78)

1. skull 2. eyebrows
3. eyelids 4. eyelashes
5. tear duct 6. frontal bone
7. sclera 8. iris
9. pupil 10. cheekbone

Just Like a Camera (page 79)

1. lens 2. lens
3. used to focus 4. iris
5. iris
6. adjusts the amount of entering light
7. film 8. retina
9. light sensitive material that records an image

Eyesight and Shape (page 80)

1. lens 2. nearsighted vision
3. corrected nearsighted vision 4. concave lens
5. normal vision 6. farsighted vision
7. corrected farsighted vision 8. convex lens

Ear and Eye Review (page 81)

1. malleus 2. auricle
3. eardrum 4. incus
5. semicircular canals 6. auditory nerve
7. cochlea 8. stapes
9. cornea 10. pupil
11. iris 12. sclera
13. lens 14. optic nerve
15. retina

The Organ of Taste (page 82)

1. circumvalliate papillae 2. fungiform papillae
3. palatine tonsil 4. lingual tonsil
5. foliate papillae 6. filiform papillae
7. foliate papillae 8. circumvalliate papillae
9. filiform papillae 10. fungiform papillae

Taste Buds (page 83)

1. brain 2. nerve
3. taste bud 4. tongue
5. taste pore 6. microvilli
7. taste cells 8. connective tissue
9. epithelium 10. nerve fiber

How Does That Taste? (page 84)

1. bitter 2. sour 3. salty
4. sweet 5. sweet 6. umami
7. salty 8. sour 9. bitter

The Nose and Smell (page 85)

1. receptor cells 2. frontal sinus
3. cartilage 4. nasal passage
5. nostril 6. teeth
7. brain 8. olfactory nerve
9. nasopharynx 10. tonsil
11. sphenoidal sinus 12. tongue

Making Sense of Smell (page 86)

Answers will vary.

A Smelly Process (page 87)

1. odor molecules 2. chemicals
3. olfactory receptors 4. scent
5. cilia 6. mucus
7. head cold 8. brain

The Organ of Touch (page 88)

1. epidermis 2. dermis
3. fat layer 4. blood vessel
5. hair 6. pore
7. nerve 8. hair muscle
9. sweat gland 10. fat cells
11. oil gland

Blemish, Pimple, Zit (page 89)

1. follicle becomes clogged 2. sebum gathers
3. a pimple forms 4. hair
5. epidermis 6. pimple
7. sebum 8. blockage
9. hair follicle 10. sebaceous gland
11. pus

Too Hot, Too Cold (page 90)

1. cold day
2. warm day
3. open sweat pore
4. goose bump
5. sweat
6. closed sweat pore
7. blood vessel
8. contracted muscle
9. relaxed muscle
10. sweat gland

Fingerprints (page 91)

Answers will vary.

Toenails and Fingernails (page 92)

1. nail bed
2. nail body
3. fatty tissue
4. bone
5. free edge
6. lunula
7. nail body
8. nail root
9. bone
10. nail bed
11. fatty tissue
12. blood vessel

Parts of a Nail (page 93)

1. nail body
2. free edge
3. blood vessels
4. lunula
5. bone
6. nail root
7. nail bed
8. fatty tissue
9. nail root
10. lunula
11. nail body
12. free edge

The Sense of Touch (page 94)

1. dermis
2. spinal cord
3. brain
4. sensitive areas
5. tongue
6. fingers
7. touch receptors
8. pain receptors
9. least sensitive

The Reproductive System: Female (page 95)

1. Fallopian tube
2. uterus
3. vagina
4. ovary
5. cervix

The Reproductive System: Male (page 96)

1. bladder
2. penis
3. urethra
4. sperm tube
5. scrotum
6. testis

Role of Reproductive Organs and Hormones (page 97)

1. testis
2. vagina
3. uterus
4. penis
5. ovary
6. progesterone
7. testosterone
8. estrogen
9. urethra

The Development of New Life (page 98)

1. develops tiny arm and leg buds; heart begins to beat
2. ears, eyes, nose, fingers, and toes are formed
3. has recognizable human features; gender can be determined
4. first movement is felt and heartbeat can be heard with a stethoscope
5. can survive birth with special care
6. fully developed with organs that can function on their own

Birth of a Baby (page 99)

1. birth canal
2. uterus
3. umbilical cord
4. placenta

Inherited Traits (page 100)

Answers will vary.

Dominant or Recessive? (page 101)

Dominant Trait

dark-colored eyes
dimples
can roll tongue
dark hair
ear points
freckles
free earlobes
hair on middle of fingers
widow's peak
turned-up nose

Recessive Trait

blond hair
can fold tongue
red hair
bent little finger
albinism
hemophilia
sickle-cell anemia

Reproductive Review (page 102)

1. Fallopian tube
2. uterus
3. vagina
4. ovary
5. cervix
6. bladder
7. penis
8. urethra
9. sperm tube
10. scrotum
11. testis

Body Nutrients (page 103)

1. protein
2. water
3. mineral
4. carbohydrate
5. vitamin
6. fat

Vitamins and Minerals (page 104)

1. vitamin C
2. calcium
3. iodine
4. vitamin E
5. vitamin D
6. niacin
7. potassium
8. iron
9. magnesium
10. vitamin A

Preventing Illness (page 105)

1. polio
2. virus
3. immunization
4. smallpox
5. bacteria
6. white blood cells
7. allergy
8. antigen
9. vaccine
10. antibody

Pressure Points (page 106)

(see page 128)

The Heimlich Maneuver (page 107)

1. food lodged in trachea
2. base of rib cage
3. fist
4. navel
5. trachea
6. trachea
7. fist
8. thumb
9. navel
10. thrust
11. lying down

All Body Systems (page 108)

1. respiratory
2. digestive
3. skeletal
4. endocrine
5. urinary
6. muscular
7. nervous
8. lymphatic
9. circulatory

Identify the Systems (page 109)

1. skeletal
2. lymphatic
3. digestive
4. reproductive
5. urinary
6. muscular
7. nervous
8. respiratory
9. endocrine
10. circulatory

Classify the Parts (page 110)

Circulatory
aorta
red blood cell
ventricle
capillary

Lymphatic
spleen
thymus gland
lymph node
tonsils

Digestive
duodenum
colon
stomach
esophagus

Muscular
deltoid
biceps
tendon
trapezius

Respiratory
alveoli
trachea
diaphragm
pleura

Sensory
malleus
auricle
retina
pupil

Body Parts (page 111)

1. nose
2. chest
3. abdomen
4. hip
5. thumb
6. thigh
7. shin
8. heel
9. forehead
10. cheek
11. shoulder
12. forearm
13. palm
14. calf
15. sole
16. instep

Scientific Sections (page 112)

1. superior
2. anterior
3. inferior
4. distal
5. proximal
6. posterior
7. inferior
8. distal
9. superior
10. anterior

Functions of Organs (page 113)

1. colon
2. heart
3. skull
4. pancreas
5. lung
6. spleen
7. bladder
8. esophagus
9. medulla
10. kidney

Grouping by Function (page 114)

Control Group
nervous system
sensory system
endocrine system

Movement Group
skeletal system
muscular system

Energy Group
digestive system
respiratory system
circulatory system
urinary system
lymphatic system

Human Body Review (page 115)

1. excretory
2. testes
3. enamel
4. epidermis
5. tendons
6. fracture
7. pressure point
8. ovaries
9. genes
10. palm
11. cervix
12. biceps
13. pancreas
14. pulse
15. pelvis
16. sebum
17. heart
18. bladder
19. pituitary
20. adrenal

Common Names for Fancy Words (page 9)

1. skull
2. jawbone
3. backbone
4. shinbone
5. kneecap
6. hip bones
7. shoulder blade
8. collarbone
9. tailbone
10. breastbone
11. thighbone

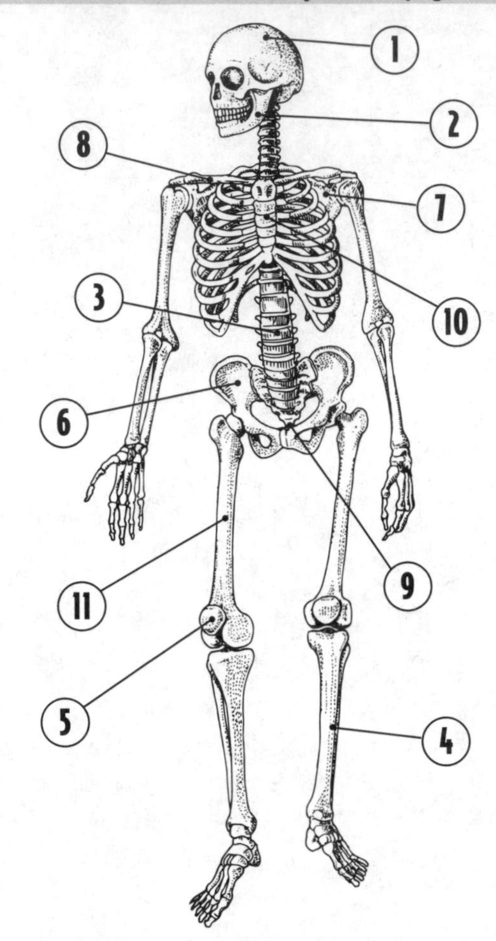

Bones of the Hands and Feet (page 10)

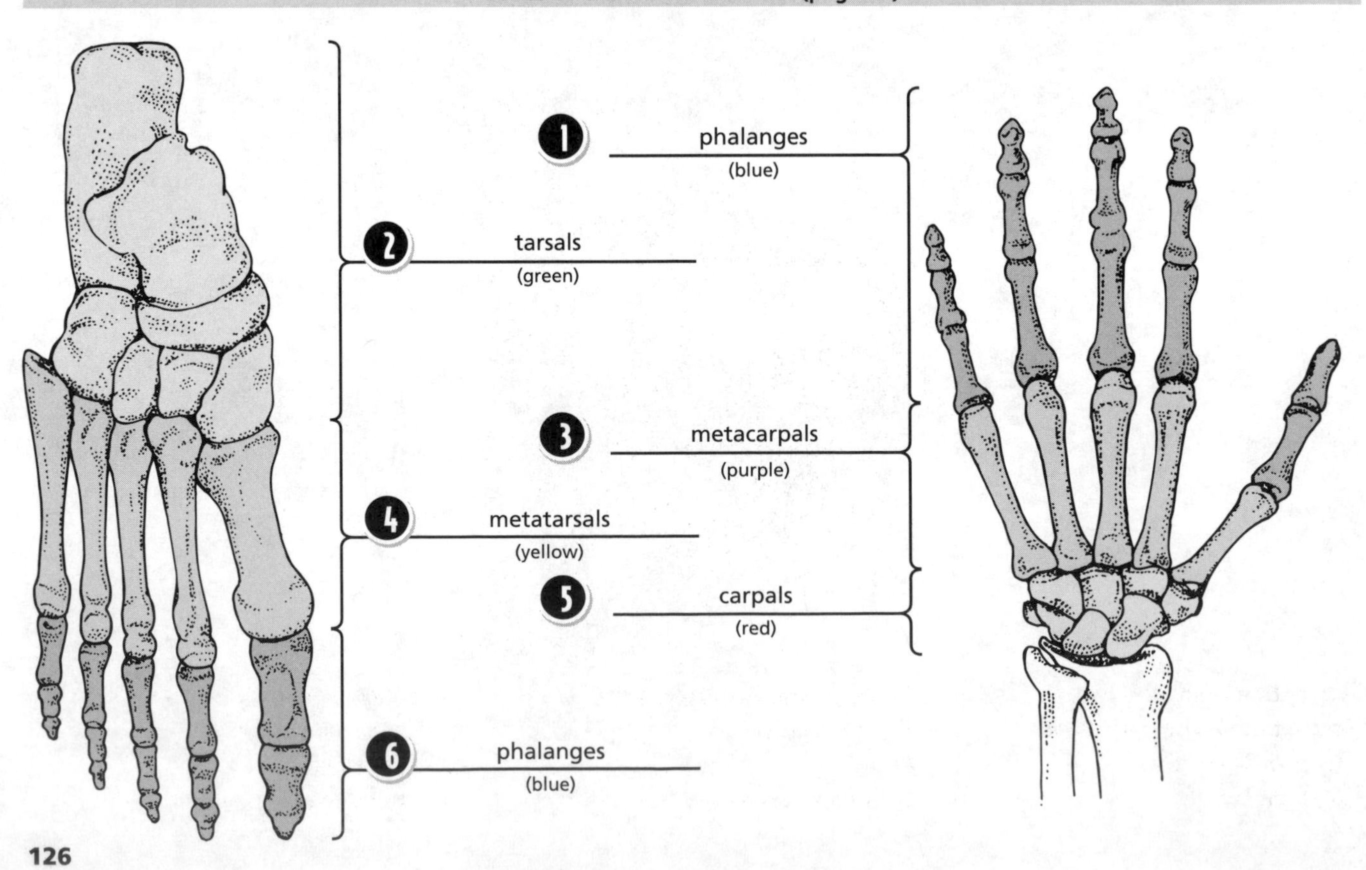

Veins and Arteries (page 28)

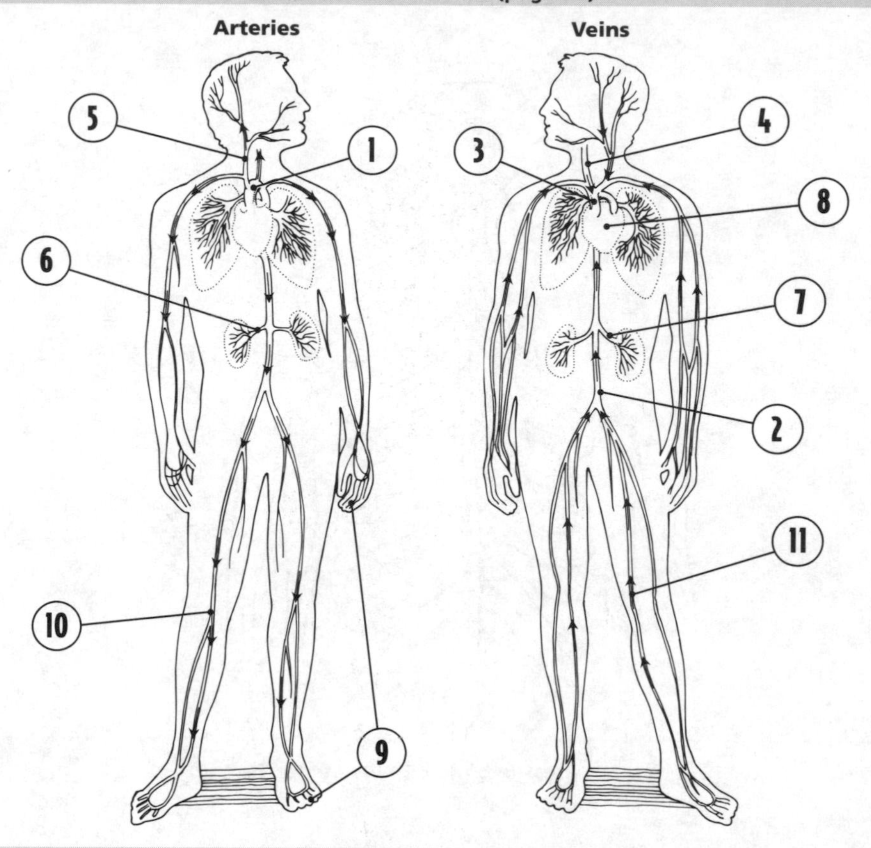

Go with the Flow, Part II (page 33)

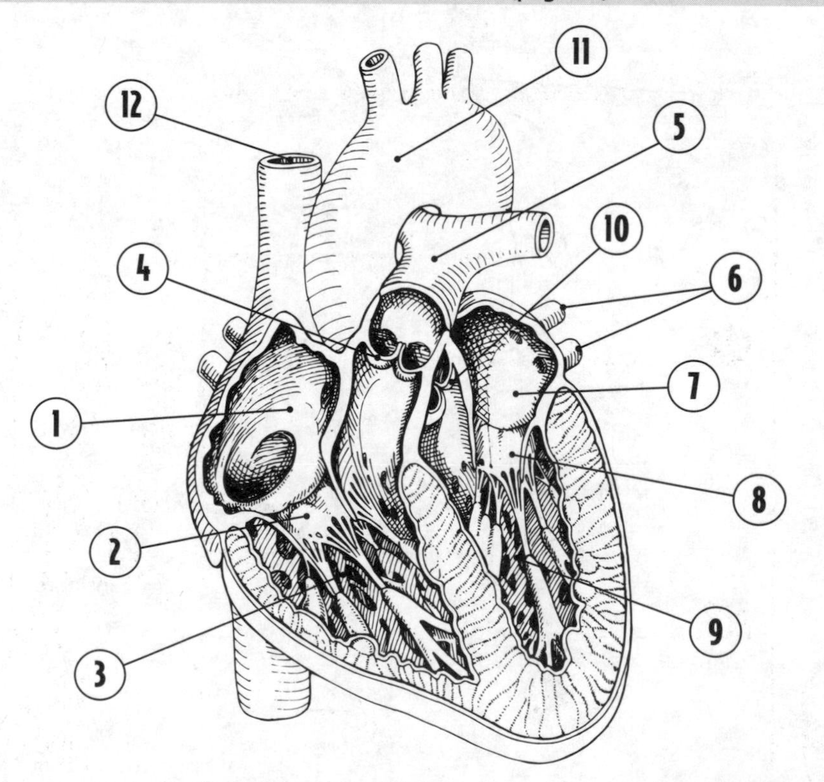

__1__ right atrium
__8__ left atrioventricular valve
__6__ pulmonary vein
__3__ right ventricle
__2__ right atrioventricular valve
__9__ left ventricle
__11__ aorta
__12__ inferior vena cava
__5__ pulmonary artery
__10__ left semilunar valve
__7__ left atrium
__4__ right semilunar valve

Nervous System Functions (page 64)

1. cerebellum
2. medulla
3. cerebrum
4. spinal nerves
5. spinal cord

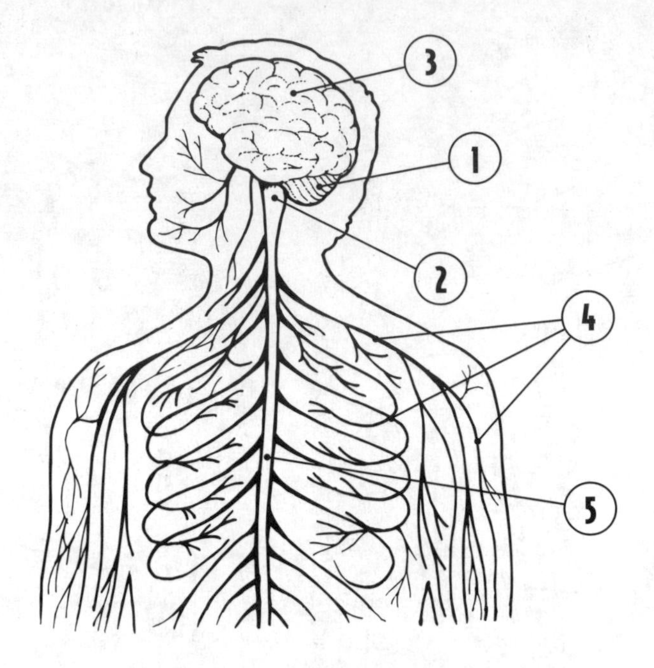

Pressure Points (page 106)

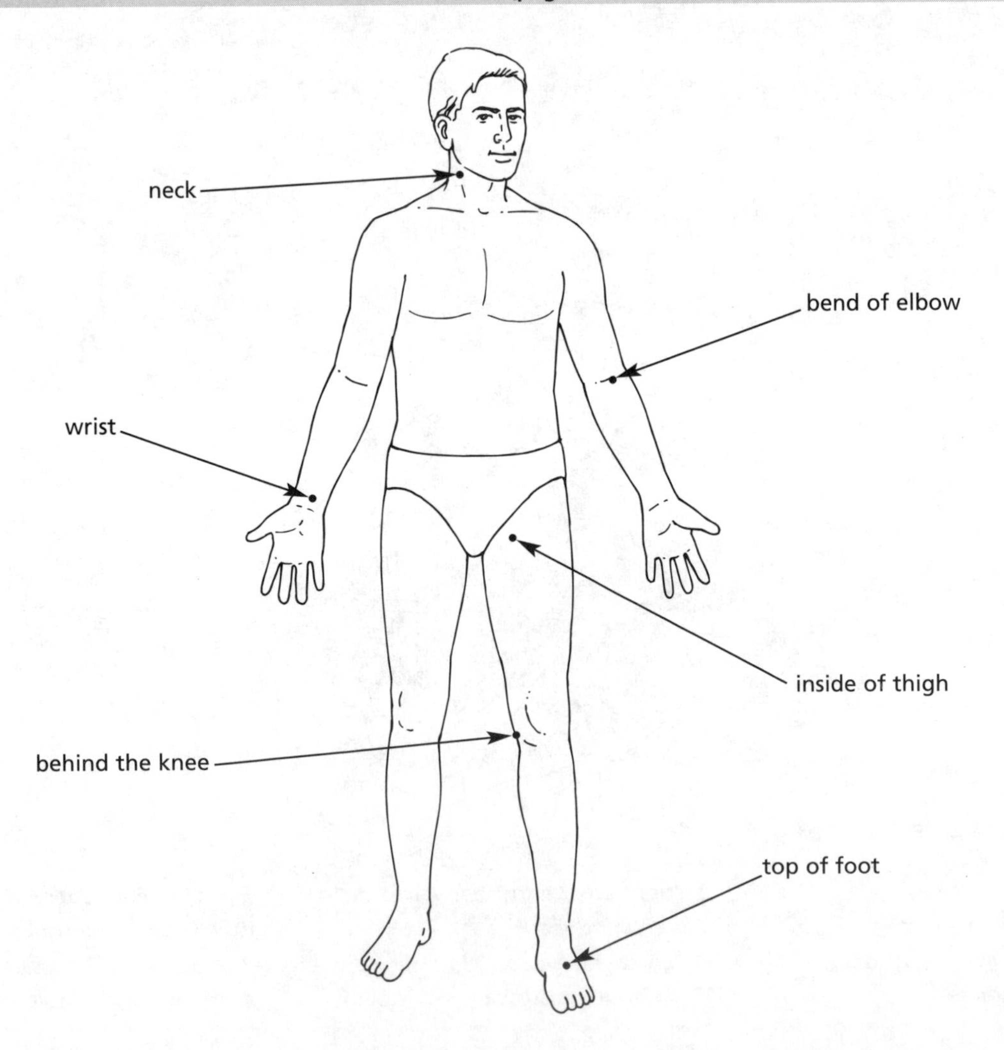